Adolf Bleichert & Co.'s

Wire Rope Systems

Peter von Bleichert

Cover by Marine Kazaryan

Copyright 2019
Peter von Bleichert
All Rights Reserved

For E.

'Flying boulder' company logo.

TABLE OF CONTENTS

INTRODUCTION 1

WIRE ROPE DYNASTY 5

CHILECITO, ARGENTINA 23

THIO, NEW CALDEONIA 49

MKUMBARA, TANZANIA 81

ZUGSPITZEBAHN, AUSTRIA/GERMANY 101

PREDIGTSTUHLBAHN, GERMANY 127

TABLE MOUNTAIN AERIAL CABLEWAY, SOUTH AFRICA 145

END OF AN ERA 161

SOURCES 167

INTRODUCTION

Each and every day, our lives are affected by the simple technology of wire rope—individual steel wire strands laid into larger bundles that exhibit immense strength and flexibility. Often called 'cables,' wire ropes suspend bridges, lift elevators, hoist materials, arrest planes landing upon naval aircraft carriers, and haul people and goods up and over mountains and other challenging terrain.

We have all ridden on wire ropeways. We call them such things as 'cable car,' 'gondola,' 'ski lift,' and 'tramway.' Like much technology that pervades modern life, the significance of the aerial ropeway is often taken for granted.

Ropeways originated in the rugged terrain of Asia: modern-day China, India, and Japan. As far back as 250 BC, it is written that people used ropeways to cross ravines and raging rivers. These ropeways comprised a track rope slung between two fixed points. The track rope allowed the load to be lightened and permitted stops for rest, even if dangling over deadly natural features. Carriers made of braided grass, sticks, and wood planks were propelled by gravity along the down-rope journey, and then pulled back up the track rope by a hauling rope powered by beast or manpower. This dual or bi-rope system, comprised of track and hauling rope, became the standard design of aerial ropeways.

Just such a two-rope system was featured in the *Taiheiki*, a Japanese epic written during the late 14th century. The *Taiheiki* speaks of a Japanese emperor stuck in a valley surrounded by enemy forces. The emperor escaped the valley by means of an aerial ropeway, crossing terrain otherwise deemed impossible to traverse. The historical record then shifts to other parts of the world.

A first European mention of ropeways came in the 1405 weapon's catalog *Bellifortis*. In South America there are references to aerial ropeways for transportation of gold ore appear as early as 1536. As ropeway design grew more sophisticated, the 17th century saw refinements, such as the one articulated by Venetian Fausto Veranzio in *Machinae novae*. Veranzio's 'new machine' comprised a wooden carrier which travelled over a fixed track rope pulled along by a hauling rope looped between two fixed wheels. It was not until 1644, however, that Dutchman Wybe Adam erected a successful large-scale operational aerial ropeway system that, arguably, was the first industrial aerial ropeway.

However, despite the ingenious innovations of these early ropeways, aerial ropeway systems were limited by the strength of the hemp-based ropes upon which they relied. It was not until the introduction of iron and then steel wire rope that the full potential of aerial ropeways was unleashed.

It was during the late 19th century—a era of rapid technology advances and industrialization—that a German engineer came of age, and saw the full potential of aerial wire ropeways: Adolf Bleichert. This is the story of his company and some of its greatest engineering achievements.

WIRE ROPE DYNASTY

Son of August Bleichert (1807-1870), Adolf Hermann was born on May 31, 1845 in Dessau, Germany. Pursuing an early interest in engineering, he studied mechanical engineering at the vocational academy in Berlin, forerunner of today's Berlin Technical University. These studies were followed by a job and practical experience with Martinischen Maschinenfabrik in Bitterfield. In this position, Bleichert wrestled with the problems of conveyance systems of the day.

In Germany's Harz Mountains, hemp ropes were used to hoist silver up through vertical mine shafts. In the damp mine environment, the hemp rotted, demanding innovation. William Albert, a mining official, invented what came to be called 'Albert Ropes:' lengths of wrought-iron wire wound and bound by smaller wire. Adolf Bleichert believed the new wire rope could be applied to aerial ropeways for the bulk haulage of goods and materials.

In 1872, Bleichert designed his first aerial wire ropeway for a paraffin manufacturer in Saxony· the southeastern German state adjacent to Bavaria and bordering the Czech Republic and Poland. Saxony's largest cities are Chemnitz, Dresden, Leipzig, and Zwickau. By 1874, the 2,242-foot long paraffin wire ropeway was hauling 13 tons per hour. This wire ropeway established the first use of iron running and guidance rails, and a two-rope configuration whereby separate hauling and track ropes were used. The hauling rope was attached to the individual cargo carriers, pulling them along a heavier fixed track rope.

Adolf Bleichert

Bleichert two-rope system: The top rope is the track, the bottom rope hauls/brakes the carrier and its load.

Instead of the large payloads transported in larger intervals, Bleichert decided upon uninterrupted conveyance of smaller payloads in shorter intervals, and arranged for double tracks, with one track loaded and the other side empty, each track travelling in opposite directions. The hauling rope moved constantly and was powered by a special drive unit, and, instead of using special rollers or supports along the ropeway, the hauling rope was supported and kept aligned by the regularly spaced carriers. Carriers grabbed the hauling rope by means of a clutch, and could release it when needed, such as in stations to drop or pick-up loads.

At the Bitterfield brickworks of friend Eduard Brandt, Bleichert built a facility to test a new aerial wire ropeway design that encompassed lessons learned with the 1872 system. This new 'Teutschenthaler aerial wire ropeway' was built and tested in May 1874. With help from fellow engineer Theodor Otto, Bleichert tuned the system, and then collaborated with Otto to create a separate concern for the construction of aerial ropeways. On July 1, 1874, the firm *Bleichert & Otto, Civil Engineers*, was founded. Engineering expertise was the initial product utilizing components acquired from other manufacturers.

During the summer of 1876, Bleichert parted from Otto and continued on his own to pursue ideas for innovations and inventions under the newly incorporated firm *Adolf Bleichert & Co*. Otto went on to start his own company with permission to use prior designs within Germany. Bleichert would regret this as he came up against the 'System Otto,' though his old partner's firm went bankrupt sometime later.

In 1877, Bleichert rented a small factory in Leipzig-Neuschoenfeld, gathering 20 workers for the shop and another six for the technical office. Here he would begin the design, manufacture, and sale of system components. In October 1877,

Bleichert brought his brother-in-law on board, the businessman Heinrich Piel.

The company won a large order from industrialist Alfred Krupp; for his factory in Sayn on the Lahn River. The installed system caught the eye of German industries and *Adolf Bleichert & Co.* quickly began to dominate the conveyance market. Many new improvements were incorporated into early aerial wire ropeway systems such as a new locking device for track and hauling ropes, as well as curve (deflection) stations for reliable and smooth guidance of carriages through ropeway direction changes.

Curve station.

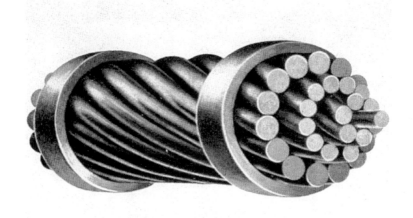

Bleichert's spiral hauling rope (originally had a hemp core)

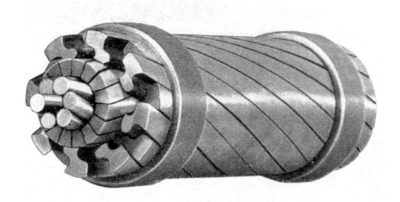

Bleichert's locked track rope

Invention and Expansion

With orders pouring in from the coal and steel industry of Rhineland/Westphalia —18 in 1878 alone—*Adolf Bleichert & Co.* increased its staff to nearly 100 employees. To accommodate expansion, the company relocated to Gohlis, a town on the outskirts of the Saxon city of Leipzig. The new factory grew beside the city's main rail line which facilitated transport of factory inputs and outputs.

On September 23, 1877, Adolf Bleichert registered Patent No. 2934 for an 'Improved Wire Ropeway System.' Under this patent, Bleichert received protection for a track rope line tensioning device that utilized chain and weights; grooved channel bearings for track rope towers; tin compound casting for connection of wire rope to chain; tilting carriers; a carrier braking system; a tensioning station for hauling ropes that consisted of a sled-mounted wheel connected to a weight on an adjustable chain; a device for automatic transfer of carriages from track rope to solid rail at stations; a curve station that allowed change of direction of both the track and hauling ropes; and, friction clutches for engaging and disengaging the carrier from the hauling rope.

The company's carrier clutches included the 'Automat' and 'Zenith' types. The *Automat* was the standard clutch, and the 'Zenith' was designed specifically for a hauling rope that travelled above the track rope. Both clutch types offered gripping that increased with loads and incline angles, and—in order to reduce jerking and undo wear and tear on ropes—they would engage the hauling rope after the carrier was accelerated to line speed on angled rails. To disengage from the hauling rope, a carrier was lifted by solid rails, thereby reducing the weight of the carrier and causing the clutch to reduce its grip and release the rope.

With new facilities and advanced manufacturing equipment, by the end of the nineteenth century, Bleichert had registered over 50 domestic and foreign patents.

Adolf Bleichert & Co.'s Leipzig-Gohlis factory.

The Land of Opportunity

Near the end of the nineteenth century, the Franco-Prussian War was won and French reparation payments of 4 billion German Marks flushed the economy. In addition, all internal borders in Germany came down, creating an internal free market (Zollverein) that saw great expansion and wealth generation. By 1880, growth slowed, though the success in Europe of the Bleichert two-rope system was attracting interest from the booming American mining industry. A representative of the company arrived in New York to publicize the system, and, in 1888, New York-based *Cooper, Hewitt & Co.*—owner and operator of *Trenton Iron Works*—received a license to build and distribute the Bleichert system in the United States.

With the release of its 1888 catalog, and in response to an advertising campaign in American mining journals, *Trenton Iron Works* booked several firm contracts for the system.

With many Bleichert Systems operating in the eastern half of the United States, *Trenton Iron Works* entered the coveted Rocky Mountains with an 1890 system built in Aspen, Colorado. Here the company installed two wire ropeways: one for transport of silver-ore, and another as a people mover for the local tram company. Bleichert Systems were soon everywhere in America, even operating as far north as on Alaska's Skagway-Chilkoot Pass route.

For the 1893 Columbian Exposition (World Fair), *Trenton Iron Works* built an operational Bleichert System. Also, at the Exposition, *Pohlig Co.* of Köln, Germany, offered what it called the 'Otto System.' This system was based on the original designs born of the Bleichert/Otto partnership.

From 1900-1920, hundreds more Bleichert Systems were installed and operated on the American continent. Though coming under control of *United States Steel Corp.* in July 1904, the *Trenton Iron Works* brand-name continued to be utilized by the steel giant's marketing division. However, by 1910, the division had been rebranded as *Trenton Tramway Factory*, and was listed in *United States Steel Corp.*'s catalog as a division of its *American Steel and Wire Co.*

Changing of the Guard

Adolf Bleichert died from tuberculosis at the age of 56 on July 29, 1901. He was put to rest in the Friedhof (Cemetery) of Gohlis. He had provided for the continuation of *Adolf Bleichert & Co.* as an open partnership and family enterprise. The Commercial Registry of October 12, 1901 officially listed the widow Victoria Emilie Hildegard Bleichert and their six children as 'Co-owners of a factory of wire ropeways'.

Bleichert groomed his sons Adolf <u>Max</u> (b. May 25, 1875) and Max <u>Paul</u> (b. May 14, 1877) to run the world-renowned firm, leaving them an efficient administrative organization, and a distribution network that reached far beyond Germany. In order to prepare for taking over the company, Max attended Koenig-Albert-Gymnasium in Leipzig and the Polytechnic in Dresden and Karlsruhe, interned in Bremen, Germany, and worked at the *Trenton Iron Works*, while Paul attended secondary school in

Leipzig and a three-year internship with an export company in Bremen, during which he conducted business in Brussels, Columbia, Mexico, New York, and Paris. With this training, Max was to manage the technical aspects, and Paul, the commercial ones.

Max Bleichert

Paul Bleichert

Max and Paul introduced gantry cranes to the product line, as well as a public relations office to take care of the 150 active patents and to publicize the company products through publications and lectures.

Offering a product portfolio of cranes, conveyor systems, electric trucks and locomotives, overhead monorail tramways, wire rope, wire ropeways, and accessories such as rope grease Diaporin, the company was able to respond to global economic downturns while expanding business.

In 1909, *Adolf Bleichert & Co.* acted as prime contractor for the design, manufacture, and sub-contracting of an entire power plant for the German city of Düsseldorf, including all intra-facility conveyance systems.

The Great War

World War I ended export opportunities, though created an internal demand for military systems that could move men and supplies to the front line, as well as evacuate the wounded to rear aid stations and hospitals. Wire ropeways were constructed that flew beneath the cover of forest canopies, traversed impassable terrain, and operated at higher speeds than those associated with commercial operations. *Adolf Bleichert & Co.* delivered 630 ropeway systems of various design to the military, and King Friedrich August of Saxon recognized the contribution by raising Max and Paul as well as their families to nobility—adding a title (von) to the surname. Later, they were also awarded the title 'Geheimrat' (Economic Advisers) to the King of Saxony.

With the defeat of Germany, the victorious allied powers recognized the superiority of the aerial ropeway as a logistical system, adopting them while prohibiting availability to the new German Reichswehr created by the Treaty of Versailles.

In 1924, the division *Adolf Bleichert & Co. Drahtseilbahn GmbH* (Gesellschaft mit beschränkter Haftung, a company with limited liability) was established for the manufacture, sale, and operation of passenger ropeways. This new division reduced risk exposure for the mother company and promoted to the lucrative people-mover market niche. In 1928, *Bleichert-Kabelbagger GmbH* was established for a new line of wire rope cranes. These excavators replaced standard iron links with wire rope, and were targeted at the open-pit soft-coal mining industry.

A Bleichert passenger wire ropeway.

50th Anniversary

With prevalent and global economic malaise, the company saw orders drop from pre-war levels of 100 per year to around eight. By its 50th Anniversary in 1924, *Adolf Bleichert & Co.* had constructed 3,000 miles of aerial wire ropeway systems including record holders:

Longest and highest elevation: Argentina, 20.6 miles to altitude of 13,940 feet.

Length of system over water: New Caledonia, 0.6 miles.

Steepest: East Africa, 86% grade.

Highest capacity: France, 500 tones per hour.

Most northern: Norway, 79° latitude.

Most southern: Chile, 41° latitude.

The Argentinean, New Caledonian, and East African cargo systems, as well as an Austro-German passenger system (which created a new Bleichert record for highest elevation) are described in the chapters that follow.

CHILECITO, ARGENTINA

On the eastern slopes of the rugged Cordilleras Mountains of northern Argentina lay deposits of iron, gold, silver and the massive lode of copper ore at La Mejicana—named for the Mexican immigrants that had discovered it—in the Famatina Range. A railroad had pressed towards the mountains from Buenos Aires, the port capital some 800 miles to the south-west. However, in 1899, the rail surrendered to impassable terrain and terminated in the province of Rioja, in the small foothill town of Chilecito (elevation: 3,227 feet). Twenty-two miles away, and some 10,500 feet higher than the town, were La Mejicana, and, specifically, the rich Upulungos Mine. In summer, the mines could be reached in two-and-a-half days by mule. The slow dangerous procession added to the per-ton cost of transport, leaving the mines undeveloped and the town's smelters with reduced profits, and, during winter months, operations had to shut down entirely.

With the arrival of the railway in Chilecito, the *Famatina Development Co.*—a British consortium that owned most of the north-western mines—pressured the central government to improve the link to the mines. The consortium's slogan became, "No Pound Sterling without a cable railway [at Chilecito]". The Argentine government committed to underwriting a wire ropeway to be built between the railhead at Chilecito and the high-altitude ore deposits. Searching among major American, British, and European firms, the Argentineans settled on what they concluded to be the safest and most efficient design: the Bleichert System.

Authorization for construction of the wire ropeway came in the form of Law No.4208, passed by the Argentine Ministry of Public Works in November 1901. An agreement with *Adolf Bleichert & Co.* was signed on July 31, 1902 for work to begin in February of the next year. With the one million gold peso order secured, the German firm began surveying the site and designing the wire ropeway. *Adolf Bleichert & Co.*'s Chief Engineer Dietrich became the firm's point-man for the project, and work began in 1903.

Design

The Chilecito wire ropeway would be the longest ever attempted, traversing unprecedented inclines of up to 50% and temperatures ranges of tropical to below-freezing. The hourly capacity was set at 20 tons up-rope, and 40 tons down-rope. Since ore is heavy, many small carriers—one every 336 feet—would be utilized, travelling at a speed of 7.5 feet per second.

In order to maintain wire rope tension, the entire system would have to be divided into eight sections, served by seven through-stations and two terminals. At each station, the carriers would be uncoupled from the section's hauling rope and passed by rail to the next section where it would pick up the track and hauling ropes again.

Due to length differences, each section's ropes would be of varying diameter. A carrier's coupler would have to automatically adapt to the new dimension while maintaining grip at high rope angles and on ice-covered ropes. The Bleichert 'Automat' clutch would mitigate these two factors as well as accommodate normal wire rope wear (A one-inch rope would soon compress and wear by 30%).

The Chilecito-La Mejicana Wire Ropeway would be primarily an ore mover, though would be multifunctional in that it would also transport personnel, lumber, spare parts, provisions, and water. The system maintained some 640 carriers of various types, including ore carriers (550-pound-capacity), line greasers (that had a slung tank and converted carrier motion into pump action, distributing a fine layer of oil on the wire rope, helping to prevent surface and interior rust), and combination carriers that moved up to four personnel, as well as mail and water.

Line greaser.

The combination carrier moved people and water (note counter-balancing tank appendage).

Gravity would assist down-rope operations. However, in order to transport goods uphill, a station's drive power surplus might range from +75% to -20%, so engineers decided to provide a steam-powered drive for each line segment. Therefore, depending on up-rope or down-rope power surpluses or deficits, each station would provide power to hauling ropes for the up-line, the down-line, both lines, or neither line. Each drive motor would range in power from 35 to 60 horsepower.

With existing maps unreliable or incomplete, Argentine and British surveyors set out to provide initial survey information, followed by tachometric elevation counts. The entire route was then photographed for orientation, and a final survey was executed by *Adolf Bleichert & Co.*

Survey team.

Survey image and the proposed ropeway alignment.

Stations

Every station had an automatic device through which the hauling rope travelled, dipping the rope into a varnish bath, and, as it exited, evenly distributing the protective material with a ring of brushes. Communication between stations was accomplished by a telephone line that ran parallel to the ropeway with every station containing a telephone booth. Also, every 3,000 feet along the length of the ropeway, outlets allowed a line inspector to plug-in portable telephones.

Station 1, Chilecito: Adjacent to the railroad and smelters, this terminal discharge station had large chutes for ore carrier loads to dump into railroad hopper cars. Arriving carriers passed over a scale that automatically printed weights for record-keeping. Once loads were discharged, the carriers travelled around solid rail loops, and could be shunted for storage or sent up-line again. Though the station had no drive equipment, the system's track rope was anchored here, and the hauling rope was kept taught with a tensioning device made of suspended concrete blocks.

The line climbed from this station at an incline of just 5%, and travelled some 5.6 miles to the next station. Due to the length of Line 1, it was divided into four parts, each with its own tensioning device.

View down Line 1 towards Chilecito.

A tensioning device on Line 1.

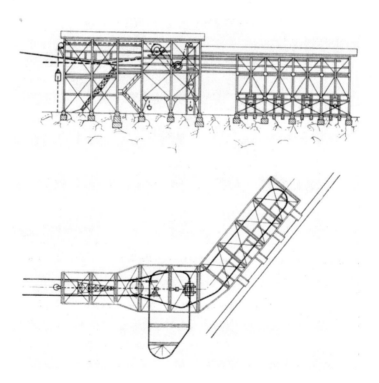

Station 1

Station 2, Durazno: This through-station was named for the slopes of Durazno y Las Higueras looming above it, though was often simply referred to as 'Mile 5.625.' The station had a drive powering the hauling ropes of both Line 1 & 2, living quarters for personnel, a shunt rail for carrier storage, and workshops.

Line 2 crossed the Amarillo River with a 1,400-foot span, and climbed at an average incline of 5% for a total of 5.3 miles.

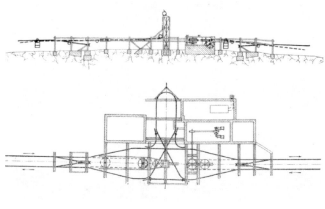

Station 2

Station 3, El Parron: Sitting on a precipitous ridge at 5,023 feet, El Parron housed living quarters.

Line 3 left the station and ran along the Cerro Alto, and climbed 1,695 feet over its 1.9 mile length at an average incline of 18.5%.

Station 3

Station 4, Siete Cuestas: Also called 'Rodeo de las Vacas,' this station has a dual drive for Lines 3 & 4, and that generated enough excess power to jump start the whole system. A station-based braking mechanism could negate this power when not needed.

Although the incline of Line 4 averaged just 5%, its 1.9 mile-length presented the greatest challenge to surveyors and engineers. Traversing the Siete Cuestas, the line passed through a 900-foot tunnel that passed carriers to solid rails and contained line tensioners at either end. The line then spanned the San Andreas canyon and skirted the steep slopes of Cerro Negro Mountain.

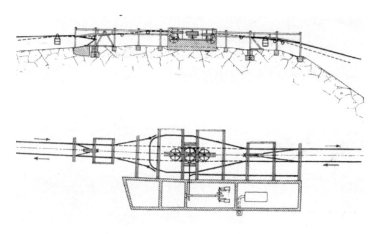

Station 4

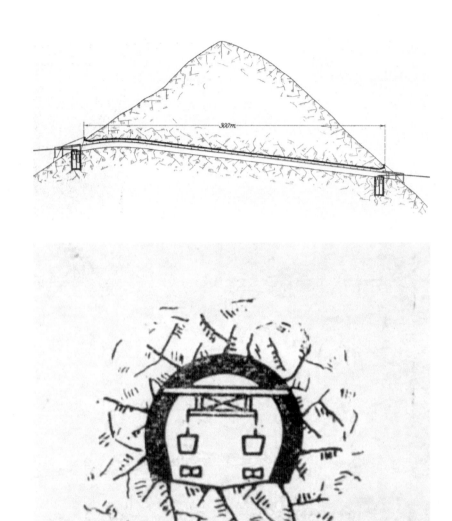

Tunnel between Station 4 & 5

Station 5, Cueva de Romero: With the wire ropeway now 14.7 miles long and pushing over 8,000 feet above sea level, a carrier arrived at Station 5 traveling about 4.8 miles-per-hour, was switched from Line 4 to 5, and passed small living quarters, as well as tensioning devices for the track and hauling ropes.

Line 5 climbed steep inclines of up to 28% along the slopes of the Cielito Rodado y Cabrera Moreno, travelling 1.2 miles.

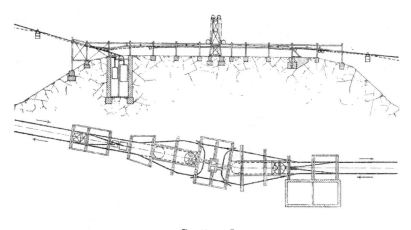

Station 5

Station 6, El Cielito: This station was equipped with a drive for Line 5, as well as a tensioning device for the 1.4-mile-long Line 6.

Line 6 spanned 2,010 feet over Rio del Rodado, then 1,725 feet over a ravine, all at inclines of up to 29%.

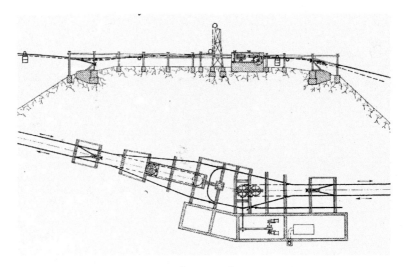

Station 6

Station 7, Calderita Nueva: At an altitude of nearly 12,000 feet, this station—also called 'Cueva de Iltanes'—powered Line 6, had living quarters, a carrier shunt rail, and a tensioning device for the 1.9-mile-long Line 7. From Station 7 up the temperature is below freezing year-round, with an average winter temperature of -4° Fahrenheit.

At a 15% incline, Line 7 leaps a 2,025-foot-high, 2,400-foot-long span of the Allada Frente a los Bayos, a heretofore unconnected divide between the mining fields and the lower range. Along Line 7, the track rope passed carriers to solid rail in order to clear a ridge on Line 7. This ridge bypass had line tensioners at either end.

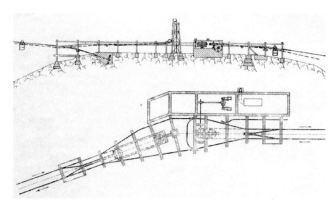

Station 7

Span at Allada Frente a Los Bayos

Ridge rail bypass between Station 7 & 8

Station 8, Bayos: This is the last through-station. The station accommodated drives for Line 7 & 8, a tensioning device for Line 8, and a carrier shunt rail. Station 8 was angled to accommodate a change in line direction.

Line 8 bounded two spans—the largest of which was 2,700 feet—before reaching the upper terminal station after 2.2 miles of travel.

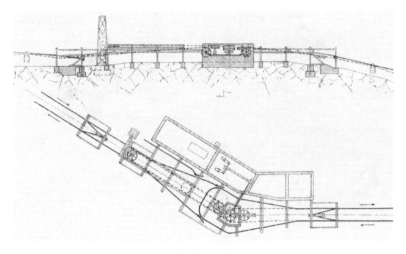

Station 8

Station 9, La Mejicana: At 13,811 feet above sea level, Station 9 was also called 'Bello Plan' for the flats upon which it was built, and was connected to the mines by a short rail line. Rail carts of ore arrived from the Upulungos mines, and were dumped into individual wire ropeway carriers for their four-hour journey down-rope to Chilecito. The track rope was anchored in concrete here, and the station had living quarters and a carrier shunt rail which held at least one personnel carrier in reserve. Water is also stored at this station in an auxiliary tank. Just beyond Station 9 was a small reversal station around for carrier turnaround and storage.

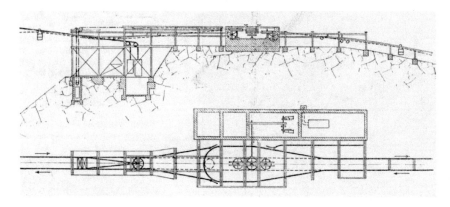

Station 9

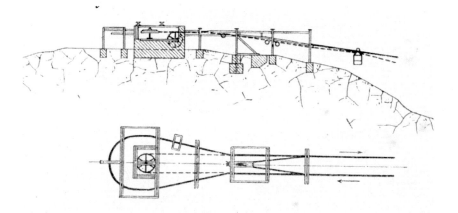

Upper reversal station

Construction

With majority of hundred-year-old mule paths deemed unsafe and unusable, the first step was to penetrate the mountains with a road that could support construction of the wire ropeway. The new road measured some 31.25 winding miles between Chilecito and the mines, and had another 37 miles of side roads.

Along the road, graders made preparations for the wire ropeway: Rugged ridges were cut out of the lime, shale, hard granite, and quartz rock, and blasting for station foundations began.

The most remarkable earth moving feat for construction of the system was the blasting of the tunnel between Stations 4 & 5. Started in December of 1903 and finished five months later, 15,000 cubic feet of rock was removed in order to allow the ropeway to pass through what otherwise would have been too large a span requiring extremely tall towers. Thirteen-and-a-half feet wide and 12 feet tall, the 900-foot-long tunnel had finished brick portals and bare rock walls.

When graders finished their work, masons constructed the system's station and tower foundations, and the erection of towers commenced.

A tower is erected in a ridge cut-out

A large compound for assembly of components was established in Chilecito. This compound was fed by the railroad that linked the site to the coast. The compound had barracks and small houses for personnel, component storage facilities, and workshops.

Vehicles were useless above Station 2, so 600 mules had the burden of transporting materials into the mountains. Therefore, no piece of equipment could exceed 330 pounds in weight. Boilers, flywheels, station and tower framing, rope wheels, and steam engines had to be dismantled for carriage, and lighter loads of water and provisions went by 90 weaker donkeys. Porters had to carry heavier loads, distributing the weight among several people. When the dangerous and difficult work slowed, in order to bring back on schedule, the Artillery of the Republic of Argentina provided 200 additional mules. At its peak, the project employed 1,200 workers, with tent camps and cooking facilities following close as the project progressed. As the ropeway climbed into the thinning atmosphere, worker wages grew and the length of their shift shortened.

Wire rope arrived on huge drums and was unrolled onto smaller spools. The wire rope for the first line was hand-carried the five-plus miles to Station 2. Thereafter, the rope spools were slung from already operating line sections, and then hand-carried again up to the next station.

Wire rope coils ready to be dispatched up-rope

Having fought blizzards, floods, and icy winds, the finish work for Station 9 was completed in December of 1904, signifying completion of the wire ropeway. On January 1, 1905, operation of the Chilecito-La Mejicana Wire Ropeway began.

Operation

The operation of the wire ropeway was managed by the Argentine Government under the Inspector General of Transportation-Division of the Inspector of Mechanics.

Line Officials and their assistants lived at the various stations. Stations with steam drive engines provided housing for a Station Manager, a Mechanic, a Boiler Stoker, a Telephone Operator, and three Line Attendants who directed incoming carriages to the outgoing rope, and walked or drove the lines daily.

Starting the ropeway was initiated by Station 9, with all other stations broadcasting in sequence that they were ready to dispatch and receive carriers. Ore carriers were then dispatched by Station 9, beginning with smaller loads and lower speeds, and then building up to full system capacity and speed.

Water carriers were stored on the shunt rail at Station 4 where there was a filling pump. Though Stations 1, 2, and 3 relied on natural springs, stations 4 through 9 had to be replenished with water.

Though the ropeway was slowed for passengers, to the unseasoned, the rapid change in altitude could bring on 'soroche':

mountain sickness. Station 3 came to be called 'Station of the Old' for those who succumbed to the thin atmosphere and had to rest there.

Conclusion

Adolf Bleichert & Co.'s Chilecito-La Mejicana Wire Ropeway of Argentina was deemed by engineers of the day as the 'Eighth Wonder of the World'. Beginning operation on January 1, 1905, the 21.4-mile system had 262 towers, over one-million rivets, miles of wire rope, and 640 carriers of all types. During the 10,584-foot climb through harsh Andean weather to the terminal at 13,940 feet, the system spanned bottomless ravines and plunged through a mountain peak tunnel. Built entirely at the Leipzig-Gohlis factory and then disassembled for transport by ship and rail, the Chilecito wire ropeway served without accidents or breakdowns until 1926, and operated until 1974 when it was abandoned in place.

Parts of the wire ropeway were re-started as recently as 2001. Though initially killing two inexperienced operators, part of the line is now open to tourists. The 'Cable Carril' is an Argentine National Historical Monument, and a small museum has been set up next to Station 1.

THIO, NEW CALDEONIA

It is the early twentieth century. A tramp steamer plies the waters off Queensland, Australia. She is headed east towards the nickel mines of New Caledonia, an island colonial possession of France. The ship is alone in the moonless tropical summer night and making way on calm seas. The breeze delivers the perfumes of land to the ship's wheelhouse. A brilliant string of stars is low on the horizon. The dark waters sparkle below them. The 'stars' are actually arc lamps that illuminate the ship loading platform and towers of the Bleichert wire ropeway at Thio.

New Caledonia was 'discovered' in 1768 by Louis Antoine de Bougainville, and was named by Captain James Cook in 1774. In 1853, France declared New Caledonia her colony. Promptly sending her worst convicts to the South Pacific territory, it was not until the 1860's that France dispatched surveyors and geologists to inventory the island. An 1864 discovery of nickel-rich surface layers on the southern half of the island—especially on the Thio Plateau and Bornet Region—piqued French interest in the island.

Mining titles were awarded by Paris to different French firms. *Société Anonyme Le Nickel* (*SALN*) was the recipient the rich areas in the south and along the Thio River valley. *SALN* moved about 1,700 employees to the island, and concentrated them around the town of Thio, the original site of a French mission.

A narrow-gauge railroad was built by the German firm *Arthur Koppel Co.*, running at 10 miles-per-hour along the Thio River to the interior. It climbed a total of 82 feet over its 9.3-mile journey. There was also a rail spur that wound its way to a smaller plateau mine to the west of Thio.

Ore deposits were mostly atop mountain peaks, and mining was achieved by open-pit terracing. Animals and porters were used to bring sacks of nickel-ore from the mines, down the steep terrain to the railroad. The ore was then transported to Thio by train, loaded on to shallow-draft barges. These barges then met larger ships anchored just off shore to transfer their cargo by hand. Overall, an expensive, dangerous, and inefficient means of exploiting the rich deposits.

Open nickel pits in the Bornet region.

The 1890 Wire Ropeway

In order to access the high Bornet mine region and connect it with the railroad that ran along the Thio River, on June 12, 1890, *SALN* ordered a 5,850-foot wire ropeway from *Adolf Bleichert & Co.*

This two-rope system had an hourly capacity of 55,000 pounds, cost the French firm 45,381 German Marks, and used Automat clutches to handle the tropical weather and typhoons that frequented the island.

Since the mineral fields were widely dispersed, a simple gravity-driven single-rope system was utilized to deliver ore sacks to a mountain-top loading station. The arriving ore was then emptied into the primary two-rope system's carriages. One hundred twenty tons of ore were dispatched from the loading station to the railway below each day. Down-rope inertia was all that was needed to run the line at 8.25 feet per second, and was regulated by an automatic brake.

In the river valley below, the discharge station incorporated a hopper capable of holding 250 tons of ore. Trains consisted of up to 15 carriages were loaded via chutes, with each carriage capable of holding 4 tons.

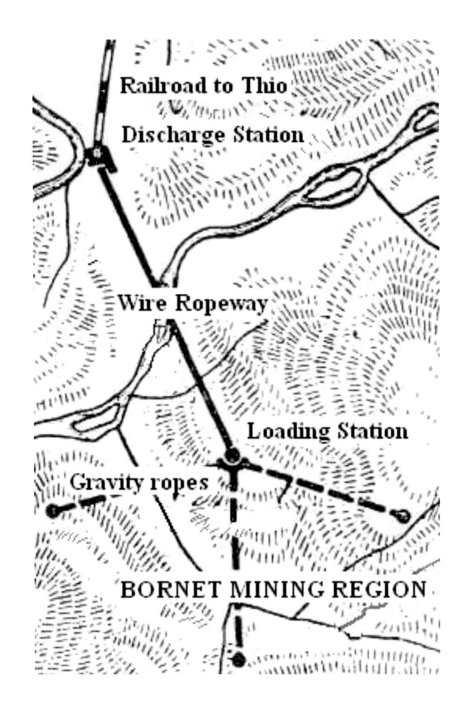

1890 wire ropeway and its location

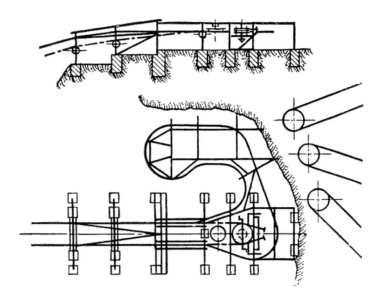

Loading station at Bornet Mines

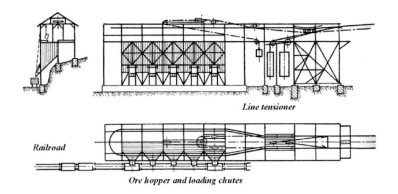

Railroad and ropeway discharge station next to River Thio.

Railroad meets ropeway at the Thio Plateau.

Sea-side Depot

Two trains per day travelled from the Bornet wire rope station to the sea-side depot at Thio where, upon arriving, the train climbed a trestle before dumping its cargo to the ground. One train per day would also service the mines at the upper fields, and bring its load of nickel-ore to dump at the depot. An excavator was used to transfer the ore to small tipper carts that were pushed out on a jetty by rail. The tipper carts were then dumped into flat-bottomed barges.

Before the steam- and sailing-ships that arrived could take on ore, they had to be towed up to two miles out to sea in order to dump sand ballast from their fore and aft holds. These ships would then return to the anchorage, and the barges—each holding between 50 and 100 tons of nickel-ore—came alongside for native crewmen to begin transferring ore by hand using baskets and buckets. The ship would then be towed out again, dump sand ballast from the amidships hold, and return to the anchorage for a final transfer. This was obviously a timely process that required calm seas.

During these operations, many storms delayed ships up to 120 days, thrashed barges about, and ripped ships from their anchorages, often depositing them on the jagged reefs. With these circumstances in mind, it was decided that something had to be done to continue the profitability of the New Caledonian mining operation.

The railroad and the 1890 wire ropeway system had made transport of ore from the mines to the sea-side depot efficient. However, it became obvious that new efficiencies had to be introduced for handling of ore between shoreline and ship.

SALN once again contacts *Adolf Bleichert & Co.* with a request for proposal.

The ore dump at Thio. Note jetty for loading barges.

De/ign

Upon request of *SALN*, *Adolf Bleichert & Co.* submitted a proposal for a ship loading platform to be erected about 0.6 miles offshore. The ship loading platform consisted of two primary structural elements; each 115 feet long and 97 feet wide, that rested on three pylons rising from the seafloor. The pylons were of stone-filled concrete; robust enough to withstand the most violent of surf conditions. Built off the peninsula that jutted beyond the breakers, the ship loading platform allowed the transfer of cargo in relatively calm waters, and was sized to moor the largest of ships alongside. It was fitted with equipment for the loading and unloading of vessels, and was connected to the land by means of a new wire ropeway. This accommodated all traffic between land and ships, including the movement of personnel.

The existing ore dump would remain and be served by a ropeway line that would connect it with the depot, a new ore transfer facility, and the ship loading platform at sea. Interconnectivity would be provided in the form of interchanges, turnarounds, and carriage storage areas.

Design of the transfer facility and ship loading platform incorporated multiple requirements: Discharge and dispose of sand ballast from ships; discharge small ore ships arriving from other New Caledonian pits; store ore on land in a dump; transfer ore from railroad to dump; transport coal to land from ships; pick up coal and load it on to ships; facilitate people and carrier traffic between land and ships; and, pick up ore and load it on to ships. These design features are represented in Figure 1.

With the proposal accepted by *SALN* at a quoted cost of 220,400 German Marks, an order is placed on August 28, 1903 for the Thio wire ropeway system. *Adolf Bleichert & Co.*'s engineers got to work.

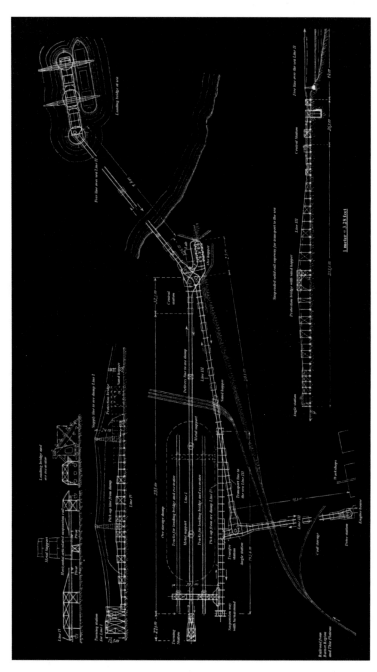

Figure 1 (view horizontally).

System drive power was centralized with all ropeway lines converging on a central station located near the sea and adjacent to the rail line from the interior. Arriving trains travelled up a small incline and approached the ore hopper (see Figure 2, "A") before discharging their cargo into the hopper's chutes.

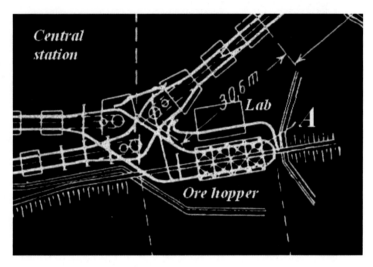

Figure 2.

Ropeway carriages from Line 1 (each carried 40 tons per hour and moved at 7 feet per second) travelled beneath the hopper chutes, over tall support towers towards the turning station (see Figure 3, "B"). The carriages then tipped their load of ore on to the dump before turning and returning to the hopper.

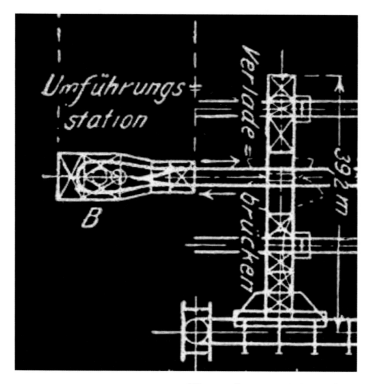

Figure 3.

Line 1 and the ore dump.

Line 1 turning station.

In order to accommodate the desired 50,000 tons of nickel-ore at the dump, the dump was raised to 65 feet high, meaning the 80-foot-tall Line 1 support towers were to be buried nearly to their tops. The weight of the surrounding ore would exert large loads on these towers, necessitating a conical sheet-metal design instead of the standard open frame girder type.

Ore could also be dumped by the train at the hopper (Figure 2, "A") and travel directly to waiting ships via central station to Line 2. Line 2 was anchored at central station, tensioned at the ship loading platform, and moved 100 tons per hour at 7 feet per second.

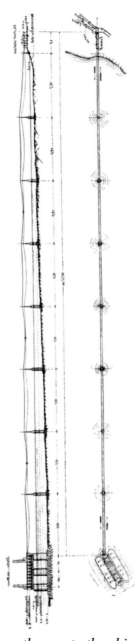

Line 2 reaches over the sea to the ship loading platform (view horizontally).

Two excavators travelled the length of the ore dump on parallel sets of tracks. Ore was scooped up and loaded onto Line 4 whose carriers hauled 100 tons an hour at 26 feet per second, were shunted to Line 3 where they are slowed to 7 feet per second, and then out to the ship loading platform via Line 2.

Line 3 served several functions: A connection between Line 4 and the sea, the connection of a 1,500-ton coal bunker and the sea; and transport of sand ballast from ships to a hopper. As ore carriers arrived at the ship loading platform via Line 2, they were dumped and refilled with sand ballast from the ships. This sand was then delivered to the hopper where it was dumped into railroad cars waiting below, taken away for dumping or fill. Line 3 also powers the entire complex with its steam generator and electric drive, enough for 100 tons an hour.

Line 4 runs 8 feet above the ground next to the dump and then climbs to some 33 feet in the air. All carriers then automatically return to the ore hopper for loading.

The loader pivots to clear the conical support towers.

All lines utilized ½-inch hauling and 1 ½-inch track ropes. When carriers were transferred from wire rope to solid tracks, double-head rails (16-inch-high; 1 ½ inch head-width) were used and suspended with cast iron clamps from portal-like structures of appropriate height.

To allow the various positions to speak by telephone, a communication wire was strung between central station, the ship loading platform, and various plug-in terminals located around the system. Since the wire ropeway operated at night, it was fully lit. Electricity was generated at the central station with excess steam from the system's power drive, and double carbon arc lamps were installed on Line 2's over-water support towers, as well as flood lights on the ship loading platform. The ship loading platform and Line 2 were fed electricity by two free-hanging tensioned insulated cables. This made the system visible to ships far out at sea.

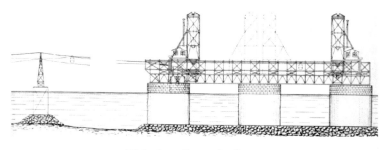

Ship loading platform.

There was a balcony-like structure at the end of the ship landing platform. It faced the land and held the rope shoes and tensioning weights for Line 2. The tensioning weights were supplemented by a flat-braid stretching rope wound over cast-iron rollers. Ore carriers arrived and were transferred from the track rope to rigid overhead rails, then pushed by hand through several switches and go-around tracks.

In order to avoid excessive lateral loads on the ship-loading platform, ships would moor using a series of anchored buoys arrayed in a circle around it.

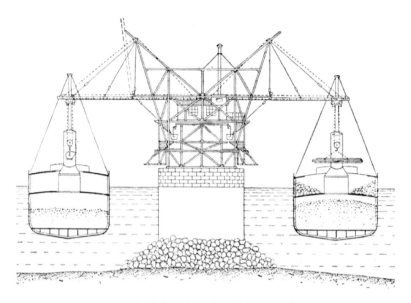

Detail of the ship-loading platform.

The two cranes—also manufactured by *Adolf Bleichert & Co.*—were double cantilever types. They had lift jibs and two-rope traveling trolleys that allowed lifting, lowering, and stopping at any point. The two cranes straddled the bridge structure and rode steel rails. The huge cranes towered some 100 feet above sea-level, and were cross-braced to resist strong winds. The wheel base of each crane unit was some 33 feet wide to distribute twisting and racking loads. For typhoons, the two cranes would be linked together to provide monolithic resistance.

Each crane had a small power shack with a steam engine and boiler. The boilers were fed by fresh water from shore, brought out by special tankard carriers. Steam was used to avoid costly undersea power cables and for ease-of-maintenance purposes. The cranes were designed to use both buckets and grippers, and travelled along the bridge by means of a three-drum winch powered by the crane boilers.

To facilitate ship loading/unloading operations, each crane was equipped with storage and discharge hoppers. As ore carriers arrived from shore, they were tipped into a 6 ½ cubic yard discharge hopper, where the crane's buckets scooped up 4 cubic yards of ore for transfer to a ship hold. Cargo going ashore—such as sand ballast or coal—was lifted with grippers or buckets and raised to a10 ½ cubic yard hopper. Chutes from this hopper filled carriers that passed beneath this hopper.

All chutes and hoppers were controlled from the crane control shacks. These shacks also housed all crane safety and emergency devices, including crane and trolley indicator instruments, bells for communication between the various shacks, and blow horns for talking to the platform's crew.

Two cranes load a small steamer at the ship-loading platform.

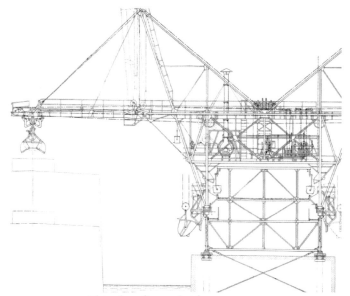

Ship loading platform crane.

Line 2 and ship loading platform superimposed on present-day Thio.

Construction

Tropical storms and typhoons lashed the site, and lightning struck steel, slowing work. Despite these challenges, the work on land went methodically, and, with precautions, safely. The greatest construction challenge was the ship loading platform.

The ship loading platform had to be far enough offshore to provide at least 33 feet of depth for deep-draft ships, requiring that the construction of the cylindrical support pylons on the muddy bottom. Realizing this was beyond their core competencies, *Adolf Bleichert & Co.* contracted with *F.H. Schmidt* of Altona, Germany to handle the masonry work.

The surging sea and muddy floor prohibited the use of *F.H. Schmidt*'s divers and their barge-mounted air compressors and support vessels. It was decided to fabricate the pylons on land, tow them in to position, and sink them in place. Dredgers removed mud from the sea bottom where the ship loading platform was to be located, reaching a layer of pebble. Upon this layer, 10 feet of quarry stones were placed, thereby compacting the foundation and preventing mud from flooding back in to the excavations. The stones also provided a firm footing for the pylons.

To form the pylons, three wooden cylinders with steel skeletons were constructed. The cylinders—some 46 feet in diameter—had a second nested cylinder with a 33-foot diameter. Concrete filled the outer ringed void between these cylinders, and the inner tube was filled with quarry stone and mortar. However, the forms first had to be floated, towed out from shore, and then sunk with accuracy.

Two small tugs tackled the job. However, despite calm seas, the pylon forms threatened to tip and swamp.

The pylon form almost capsized during tow.

Once in position, swimmers opened inlet flaps, allowing the forms to flood. When submerged and settled upon to the quarry-stone base, the forms were anchored, and concrete was pumped into the outer ring. Once set, the inner tube of the form was filled with the quarry stone and mortar.

Installing the ship mooring buoys and anchors was also a challenging feat: Large concrete blocks with chain had to be built, floated, and submerged.

The ship loading platform under construction

Supports for Line 2—the over-water portion of the ropeway—also required marine foundations. Some would be built atop rocky reefs, with others, on soft seabed. For those atop reefs, divers built walls of concrete and back-filled them with quarry-stone. For muddy or sandy bottom conditions, steel piles were driven, and foundations built upon them.

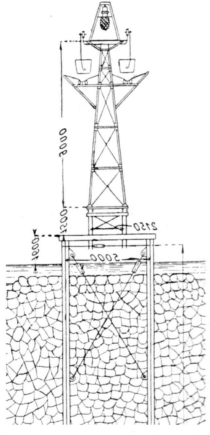

*Line 2 marine support tower for mud bottom.
Note the high-intensity lamp that made the towers visible to ships at sea.*

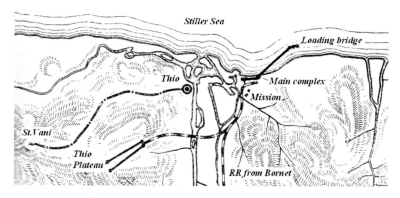

The wire ropeway and its surrounds.

Artist's impression of the unique marine ropeway.

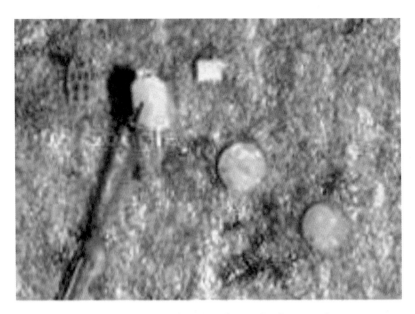

Existing to this day, the ship loading platform pylons as seen from orbit.

Operation

Several simultaneous operations could be performed by the system: two vessels could be discharged or loaded at once. With Line 2 able to handle 100 tons an hour in each direction, a total turnover of 200 tons an hour was possible. Every hour, the cranes could handle 100 tons of ore, 60 tons of sand, and 40 tons of coal, giving them a combined maximum hourly loading ability of 200 tons. Therefore, a 3,000-ton capacity vessel could now be loaded in three days versus the old record of 50 days. Also, one-dozen trained operators had replaced hundreds. On land, the lines and stations required less than 100 workers, allowing *SALN*'s workforce to continue mining during ship loading and unloading operations.

After a manufacturing time of two years, and an assembly/construction time of three years, the large and intricate system began operation in 1906, exporting some 9,000,000 pounds of high-quality nickel-ore in 1907 alone. All parts were made in the Leipzig-Gohlis factory, and shipped to the remote island from the European ports of Antwerp, Hamburg, and Le Havre. Over 2,000 tons of steel parts were required to create the Thio system.

The Typhoon of 1909

On February 11, 1909, the barque *Joliette*—her holds partially filled with cobalt—arrived at Thio to take on a load of nickel-ore for delivery to Sydney, Australia. A typhoon had followed her in, and huge swells began to tear at the island. Attempting to maneuver and keep her bow to the heavy seas, *Joliette* succumbed to a rogue tidal wave that spun her, and, as she broached in the wave trough, slammed her in to the ship loading platform. The ship loading platform's metal superstructure was torn and twisted, and the *Juliette*'s hull shattered on the stone pylons, and her mast collapsed into the sea.

The ship loading platform after the typhoon.

The wreck of the Joliette upon the ship loading platform.

Conclusion

Adolf Bleichert & Co. was not contracted to make repairs to the ship loading platform and other minor Typhoon damage to Line 2. It is therefore assumed that, though the rest of the facility remained operational, the system's ability to directly transfer ore to ships could not continue. However, on January 22, 1912, *Société Anonyme Le Nickel* placed an order with *Adolf Bleichert & Co.* for a new ropeway for the island's interior. This new system comprised a single two-rope line with a loading station at Noumea, a single angle station, and a combination discharge and drive station where it met the narrow-gauge railroad. This system was 9.46 miles long, had a capacity of 20 tons per hour, and cost 500,000 German Marks.

Though the Thio system was disassembled at some point, to this day, the pylons of the ship loading platform provide a mooring for ships. The *Eramet Group* continues mining operations on New Caledonia to this day.

MKUMBARA, TANZANIA

It is 1906 in German East Africa (modern-day Burundi, Tanzania and Rwanda). Imperial Germany's governor seeks to exploit the natural riches of the colonial territory. In the Usambara Mountains, which lie in the northeast of modern-day Tanzania—land of Kilimanjaro, Lake Victoria, and Zanzibar–grow vast stretches of tropical wood, including the valuable cedar groves of the Shume Forest (Schumewald). The Usambara Railroad is being built to follow the Pangani and then the Mkomasi Rivers inland from the Indian Ocean port city of Tanga, and onwards to the planned Mkumbara depot.

The plantation company *Wilkins & Wiese* negotiates for rights to exploit the timberlands. They plan a saw mill at New-Hornow, deep in the ancient forests. Quickly realizing that the railroad cannot traverse the steep mountains and their crumbling weathered faces, on June 8, 1906, *Adolf Bleichert & Co.* is hired to construct a wire ropeway to link the railway and the mill.

The resulting Mkumbara Wire Ropeway is 5.67 miles long, traverses the gaping Ngoha-Goatal Valley, contains unprecedented free spans, climbs 5,000 feet from the valley grasslands to the mountain forest, and, with 41° inclines, is the steepest continuously moving ropeway in existence.

Delivering timber and goods between the mountain-top saw mill and the railway depot, the system opened to much fanfare in 1909, and had a down-rope hourly capacity of over 22,000 pounds.

Part of German East-Africa (Tanzania). Areas of interest boxed.

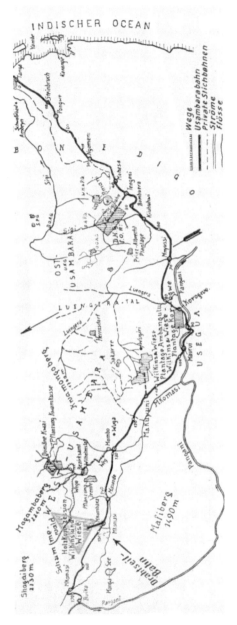

The aerial wire ropeway is in red. Note the Usambara Railroad (black line), and triangular timber harvest concession owned by Wilkins & Wiese (Outlined in green. View map horizontally).

Design

Though wood was readily available for construction, East Africa's climate and termite populations mandated the use of steel for the system's support structures. With analysis of projected loads, tension requirements, terrain, and deflection angles between Mkumbara and the saw mill at New-Hornow, company engineers concluded the system would be comprised of three lines, and that deflection angles—changes in the wire ropeway's direction—dictated two angle stations. The two-rope system had its track rope strung above the hauling rope, and would thus utilize the company's Automat clutch.

The loading station and saw mill at New-Hornow.

The saw mill works of *Wilkins & Wiese* at New-Hornow was at an altitude of 6,562 feet and located in a large timber concession of the Shume Forest, a wooded plateau of the West Usambara Mountains. The mill had four frame saws that were powered by a 90-horsepower portable steam engine. These saws could reduce all but the mightiest of trees.

Extremely large old-growth cedar and indigenous podo trees were harvested and transported to the mill. They were then cut into beams, boards, and logs before loading on to carriers. Once aboard the ropeway, cargo was hauled down to trains waiting at Mkumbara and transferred to carriages of the Usambara Railroad. The timber then travelled some 92 miles to the Port of Tanga for export. Most trees ended up as pencils manufactured in German factories.

Adolf Bleichert & Co. offered a variety of timber carriers. They were light, simple, sturdy, and had automatic lubrication. Medium and heavy loads required two four-wheeled runner assemblies, while smaller loads allowed the use of single runners. Some carriers could handle 45-foot-long 2-ton tree trunks.

Large log carrier.

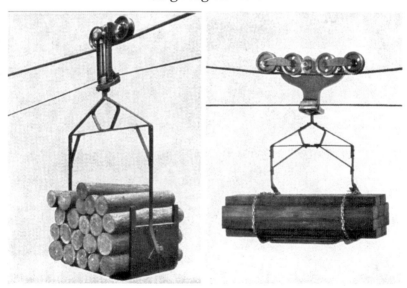

Small log and beam carrier.

Power & Braking

The system had a 75-horsepower electric motor powered by the mill's steam engine, as well as a generator, and a bank of batteries installed by *ConzElektrizitaets* of Hamburg, Germany. Excess down-rope inertia was captured, generating electricity that recharged batteries and lit the saw mill, living quarters, workshop, and stables. The electric motor was used primarily to jump-start the line, bring it to a full stop, or, in cases where up-rope demands exceeded down-rope inertia, for hauling carriers. A patented compensating gear protected both the hauling rope and motor from each other.

The system brake was an automatic hydraulic mechanism, supplemented by the main electric motor. The automatic braking system, designed by *J. Schreider* of Säckingen, Germany, utilized two reservoirs of soapy water in a box separated by a restrictive baffle. Powered by the ropeway via reduction gears, a pump moved the water from one reservoir and forced it into the next through the baffle. Should the speed of the pump increase due to heavy down-rope or light up-rope traffic, the restrictive baffle automatically closed in proportion to the increase in speed, reducing the pump action and creating a back-pressure that slowed the pump, thus slowing the ropeway and ensuring safe line-speeds. This automatic braking was supplemented by large rotating terminal wheels that were connected by belts to the electric motor.

To slow heavy down-rope loads or to bring the system to a full stop, operators engaged the motor.

Stations

The loading station at New Hornow included rope anchors and the system's machine room. The station was also equipped with fixed suspension rails and a shunt for storage and loading of empty carriers.

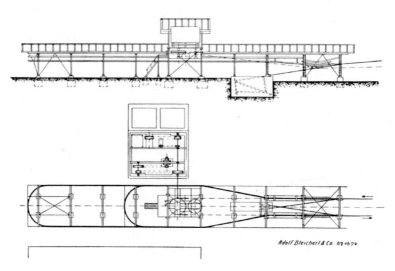

Loading station at New-Hornow. Note machine room that contained electric traction motor, batteries, the automatic braking mechanism, and water storage tanks.

The Loading Station was built adjacent to the saw mill, had a water tank on its roof, and two large cisterns for storing rainwater for use during the dry summer months. The tank and cisterns supplied the braking mechanism and the system's human operators. Timber carriers were manually dispatched from the station. Once on the line and at speed, carriers began a gentle ascent before cresting the plateau edge. When over the precipice, they plummeted some 984 vertical feet along the mist-shrouded mountains. On this first section of the line, carriers travelled 8,203 feet over 27 individual support towers.

The wire ropeway crests at 6,598 feet before its steep descent.

Carriers traveling down-rope then arrived at Angle Station 1. Located at an altitude of 5,414 feet, the station has shaft-slung weights (concrete-filled steel boxes) that maintained rope tension. As a carrier entered the station, the track rope passed it to a solid suspension rail. This rail deflected the ropeway 40° and then passed the carrier back to the track rope. Hauling power was transferred from the first line to the next through gearing. Leaving Angle Station 1, the ropeway slopes down at almost two feet for every foot travelled. This steep drop required special iron-clad steel tower rope shoes that prevented the track and hauling ropes from derailing.

Angle Station 1 clings to the mountain-side.

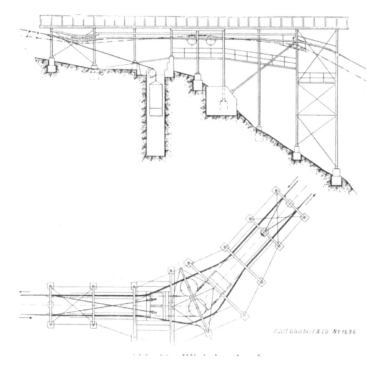

Angle Station 1.

This steep section of the line travelled a distance of some 4,000 feet over 11 towers.

Next, carriers arrived at Angle Station 2. Their runners disengaged from the track rope and transferred to solid rails to ride through a 15° deflection. This deflection set up the line for direct travel to the final station at Mkumbara Terminal. Reaching the end of Angle Station 2, carriers left rails for the track rope, and their clutches engaged the hauling rope. At this station, hauling ropes were tensioned by weighted, sled-mounted wheels. Hauling energy was passed from one section to the next through gearing.

Carriers leaving Angle Station 2 leapt over Ngoha Valley. The unsupported 3,000-foot free-span ended with a special spanning tower that maintained tension in both ropes.

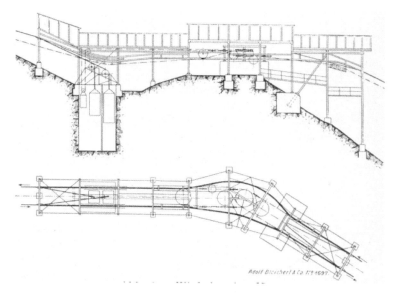

Angle Station 2.

Spanning tower between Angle Station 2 and Terminal.

This final section of the line was strung over 38 towers, including several towers that included tensioning devices. The ropeway then settled into a shallow slope, and carriers glided on to the terminal.

One hour after leaving the Loading Station, carriers and their cargo arrived at Mkumbara Terminal Station (altitude: 1,367 feet) where the wire ropeway met the railroad. Carriers transferred from the track rope to suspended rails. The rails deflected the line, aligning carriers parallel to the railroad tracks. A raised dock facilitated cargo transfer. When carriers released logs, gravity rolled them on to railroad wagons, while cut lumber was transferred manually.

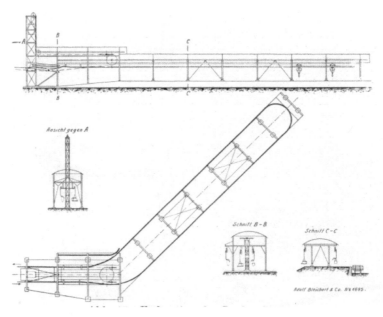

Terminal at Mkumbara.

Carriers arrive at the terminal.

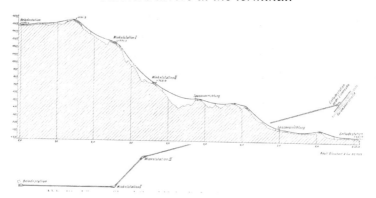

The wire ropeway's profile.

Survey photography.

Construction

It took 103 men seven months just to move the mill's steam engine to New-Hornow: It was mounted on a railroad carriage and moved forward on a short bed of track. The track was then picked up from behind the carriage to be reinstalled ahead of it. Depending on the terrain, progress was 300 to 3,000 feet a day, with the steam engine often teetering on chasm edges, and threatening to tumble away. Though independent of the primary construction contract, this Herculean effort set the tone of the whole project: Progress would be a slow arduous process of clearing pathways, making bridges, and blasting rock.

On April 16, 1907, the firm *P. Hoefinghoff* of Dar-es-Salaam was hired to build concrete foundations for the stations and towers. The contractor began work in June of that year.

With the Usambara Railroad terminating in Mombo and still being built toward Mkumbara, the transportation of supplies became problematic. About 4,000 drums of cement and 1,213,000 pounds of steel had to be hauled by horse-drawn carriages up the valley. Since Tsetse flies prevented the use of animals on mountain trails, these supplies had to be carried up the rugged mountains on the heads of natives.

With steep canyon drops and crumbling weathered stone, falling rocks slowed construction. Many retaining walls had to be built and endless face-cracks were filled with concrete. Stubborn terrain, inexperienced local workers, lack of suitable pathways, and torturous sand fleas and mosquitoes all combined to make the project extremely challenging.

To complete construction, thousands of cubic feet of rock had to be removed and the earth leveled in multiple locations. In one case, it was decided to add a support tower–Tower 23A—just to avoid excessive blasting between New-Hornow and Angle Station 1.

Operation

Small, comfortable electrified houses were installed along the system for operators and technicians. These houses and the ropeway stations were all connected by a telephone system.

Despite being a cargo-only system, passengers would on occasion ride platform carriers designed to move supplies, or ride beams, or logs. Mr. Wiese—the saw mill's owner—would often invite his entourage to ride the contraption. Among those to accept were the Colonial Lieutenant Governor as well as big-game hunters.

World War I

The Usambara Mountains emerge from the clouds, emerald green and jagged. Gliding on wire rope, a load of timber silently descended through the mist. The silence of the forest was broken by the thunder of war echoing in the distance, announcing its inevitable approach. Just five years after opening, the Mkumbara/New-Hornow ropeway was disassembled and buried, its support towers blown, all to deny the system to the advancing British forces.

Despite four years of effort and tens-of-thousands of casualties, the British were unable to capture German East Africa or subdue its defenders. Imperial German General Lettow-Vorbeck ran a successful campaign of resistance and tactical elusiveness, though, once given proof that Germany had conceded in Europe, the general surrendered his sword. However, Great Britain did seize the Usambara Railroad in 1916, securely holding the mountains to its north, including Shume Forest. When the war in Europe ended, Great Britain gained administration over German East Africa under a League of Nations mandate. Tanzania would later declare independence on July 7, 1961.

A load of boards descends to Mkumbara on the Bleichert system

Present Day

In 1985, a permanent replacement ropeway was constructed by *Wyssen Cableways* of Reichenbach, Switzerland, once again linking New Hornow—now the village of Nywelo and nicknamed 'Shume Machine' for the saw mill and 'sky-line'—and Mkumbara with wire rope. This modern system continues to transport timber to the Tembo chipboard factory and the Usambara Railroad.

It is said that if you speak nicely to the operators, one might even be allowed to ride a carrier down the mountain…at your own risk, of course.

ZUGSPITZEBAHN,
AUSTRIA/GERMANY

Münchenerhaus at the Zugspitze.

Among a series of three limestone peaks in the Wetterstein Range of the Austro-German Alps, stands Zugspitze. At 9,718 feet tall, it is the highest mountain in Germany. In its shadow, and cradled among Wetterspitze (9,429 feet), and Plattspitze (8,822 feet), lies the Schneefernerkopf, an alpine plateau of ice and snow that is a winter sports paradise.

The Zugspitze is located on the border between Austria (State of Tyrol) and Germany (Bavaria). Shared by both nations, the mountain has seen its share of political machinations and competitions to reach its heights. However, the most impressive attempts to get to the Zugspitze summit have involved wire rope.

Bleichert's Zugspitze wire ropeway

History

Cog railroads—funiculars—had opened the mountains of the world to safe exploration. These systems suffered limitations, however, as they required a steady vertical and horizontal alignment. Often steep and deep crevasses blocked the right-of-way, and required difficult and expensive foundations, bridging, and tunneling. Though practically weather-proof, funiculars did not offer the operational benefits of capacity or speed that wire ropeways did.

In 1871, Germany hired Austrian engineer Cathrein to access the peaks from their side of the border with a funicular. The Rigi-Tramway soon opened, carrying visitors to airy alpine heights via steep climbs, sharp curves, and impressive tunnels. Reaching the Zugspitze, however, proved to be too great an engineering and financial challenge at that time.

In 1897, the Münchenerhaus was built at the Zugspitze summit. This high-altitude lodge was a place for adventurers, climbers, and skiers to rest, and, by 1900, a meteorological tower was added. A roped path snaked from Münchenerhaus and along a ridge line, providing safe access to the ice-field.

In 1924, a Concession Application was submitted to the Austrian Federal Minister of Commerce and Transportation. *Austrian Seilbahn Co.* of Vienna—supported by Professor Findels of the Polytechnic of Vienna—began preparatory work for a new wire ropeway to reach the peak.

Doctor Kleiner, owner of the engineering firm *Kleiner of Innsbruck Construction Co.*, had a thorough knowledge of the Zugspitze terrain. He proposed an ambitious wire ropeway alignment leading from Obermoos near Ehrwald (3,937 feet above sea-level), past the Wienerneustadter Chalet just below the west summit of the Zugspitze, and climbing to connect the Münchenerhaus to the valley. Surveys determined that it was best to locate the mountain terminal station on the west slope of the Wetterstein range at an altitude of 9,203 feet.

World War I interrupted the 1912 plan to build a passenger wire ropeway that would open the Zugspitze summit from the German (Bavarian) side of the border.

The Austrian State of Tyrol also issued permits to open up the Zugspitze from the Austrian side of the border. In 1924, the bulk of financing was still missing from the equation. However, despite post-war economic hardship, Attorney Hermann Stern, the Vice Mayor of Reutte, and City Councilman Opitz of Berlin, were the greatest advocates for construction of the wire ropeway, and managed to put together financing. With contract provisions to support local firms and industry, the government of Tyrol underwrote the ropeway. *Zugspitzbahn Co.* was formed to represent the investors and manage the project.

The passenger division of *Adolf Bleichert & Co.* was awarded a construction contract in November of 1924. Engineers, forgers, and mechanics at the German transportation firm got to work.

Enter *Adolf Bleichert & Co.*

The Zugspitzbahn would break the company's own record for highest system, overtaking its Chilecito-La Mejicana system in Argentina.

After extensive study of the mountain, *Adolf Bleichert & Co.* proposed a design that would reach the summit from Obermoos outside of Ehrwald. Though the company had already made a study for Germany, proposing a ropeway from Lake Eibsee to the summit, World War I and its aftermath had precluded such expenditure by Berlin. Therefore, this old plan is shelved in favor of the Austrian contract.

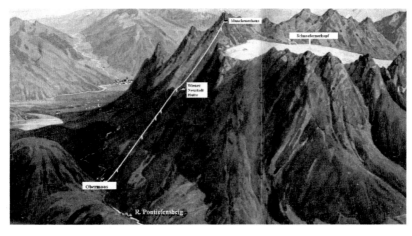

Bleichert's proposed wire ropeway alignment.
Note the Schneefernerkopf ice-field.

Seeking to diversify and grow, the Bleichert formed a dedicated passenger division in 1924 called *Adolf Bleichert & Co. Drahtseilbahnbau GmbH*. Related to its formation, Bleichert's passenger division had also entered into an alliance with Italian engineer-industrialist Luis Zuegg.

The resultant Bleichert-Zuegg System permitted sweeping rope spans between support towers, as well as increased cabin weights and speed of travel. The basic Bleichert-Zuegg System comprised two passenger cabins supported by a looped track rope and secured to each other by a looped hauling rope. The system counterbalanced the cabins, with one at the lower station when the other was at the upper station. The weight of a descending cabin maximized ascension energy for its counterpart and eased the burden on the ropeway's drive motor. The system included an auxiliary hauling rope that could take over from the primary hauling rope when needed, though otherwise remained parked.

Design

Due to climatic conditions and space limitations at the summit, unlike most Bleichert-Zuegg systems, the Zugspitzbahn's drives were in the lower station.

Adolf Bleichert & Co. engineers compiled designs, and the manufacture began in the company's two Leipzig factories: Gohlis and Eutritzsch.

A local power plant in Reutte—a major underwriter of the project—transmitted 8,500 volts of direct current to the ropeway. Power was stepped down to 220 volts with transformers at the Valley Station, and fed into a three-phase electric motor which powered a generator. The generator powered the primary and secondary drive motors, feeding them various levels of amps to allow control of drive motor speeds. The generator also charged a bank of reserve batteries. There was a primary (115kW) and secondary (50kW) electric motor system, as well as a back-up diesel (160 hp) engine. Both drives could power the main or secondary 'emergency' hauling ropes, and each had a separate electric motor. The two drives—each located on top of the other—could be mechanically coupled, allowing the secondary drive to pass its power through the main drive, or vice-versa. The electric drives put out up to 100 horsepower to handle the heaviest anticipated line-loads. These loads occurred when a full cabin ascended, and an empty cabin descended. When the descending cabin was full, the system operated at only 50 horsepower output. A full descending cabin and empty ascending one could generate and return electricity to the batteries and local power grid. If power was lost, the batteries could take over without interruption. Should both primary power and batteries be unavailable, the diesel engine stood by. The diesel could turn the main generator which would then power the drives. Should the generator also fail, the diesel engine was linked directly to the drives with leather belts.

The track rope was a 1.89-inch diameter spiral-stranded Hercules-type. For the Zugspitzbahn, this rope was manufactured in a single 11,484-foot length. The monolithic strand reduced wear-and-tear on the rope that cabin running gear and couplers would cause. The track rope spool weighed some 88,000 pounds, and was rated to carry 388,000 pounds. The load on the Zugspitzbahn was projected to be just 25% of that limit. The primary hauling rope had a diameter of 1.1 inch and was also built as one long unit. The secondary hauling rope had the system's smallest diameter at ¾ inch.

The hauling rope drive wheels of the system's two sections were linked by differential gears and sheaves, and the support towers had long shoes that broadly supported it. The hauling rope moved over a series of ball bearing-backed grooved rollers, reducing rope damage, and extending rope longevity.

There was a total of six towers designed for the Zugspitzbahn. The first tower was located 531 feet from the Valley Station and was 98.4 feet tall. Tower 2 was 3,831 feet from Tower 1 (this is a record span-length between support towers for a cabin holding up to 20 people). Tower 2 stood some 103.4 feet above the frozen rock. Tower 3 was 4,691 feet from the Valley Station. At this tower, the track rope came to within 42.5 feet of the ground. Tower 4 is 52.5 feet tall, and was perched next to an alpine hut called Wienerneustadter. Both Tower 3 and 4 were installed at the area known as the Ehrwalder Peaks. Tower 5 was 5,669 feet from the Valley Station and was 70 feet tall. From Tower 5 to 6 was a 3,281 foot free-span. When the passenger cabin was between these two towers and fully-loaded, it passed just 394 feet above the jagged limestone Alps. Clacking over Tower 6, the passenger cabin approached the upper station.

Of the two stations, the Mountain Station was more challenging to design and build. Its location was precarious, and its perch, icy-steep. Having to be partially dynamited out of solid rock, this summit station had to be able to withstand the extreme temperature and winds at altitude. The station's appointment was to be classic, rich, and comfortable, allowing indoor and outdoor vistas, as well as access to the upper ice field.

Mountain station.

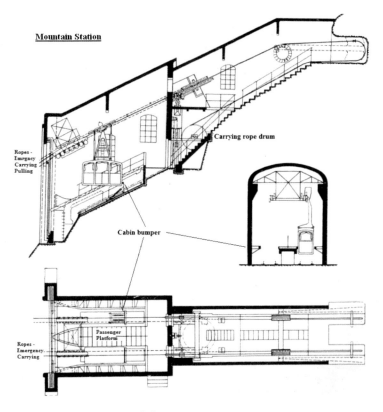

Mountain station

Appearing to be a luxurious alpine lodge, the Valley Station of the Zugspitzbahn would house the ropeway's mechanicals and passenger loading area. This lower station provided comfortable passenger waiting areas, a restaurant, sleeping accommodations for ropeway personnel, and a hotel for visitors. The Valley Station housed the hauling rope drive wheels, the ropeway tensioning devices, the drives, generator, and control room.

Valley Station/Hotel

Visitors at the Valley Station/Hotel.

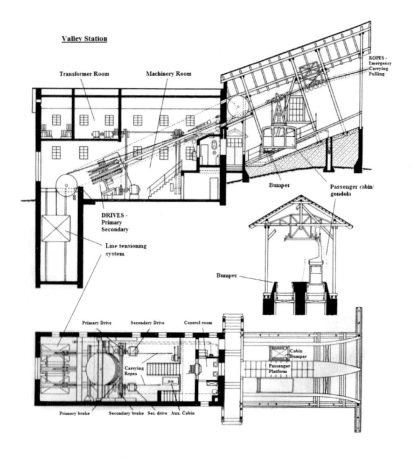

Carriers on the Zugspitzbahn had large glass- and plastic-paned windows that allowed sweeping, unobstructed views of the scenery, and could be opened when weather permitted. These passenger cabins, made primarily of aluminum, were designed to be light. This design approach facilitated the large spans between support towers on a Bleichert-Zuegg system. The cabins were hauled along the track rope at a speed up to 10 feet per second, hung freely from overhead runner assemblies, and, independent of line angles, self-leveled. The runner assembly had eight runners that distributed the cabin weight (a fully loaded cabin weighed 6,173 pounds), and had an emergency clutch that could lock on to the track rope. From a control panel next to the door, the cabin conductor operated the door, cabin brake, and telephone. Telephone communication between the cabin and stations was provided by an electric signal system conducted by the wire ropes. Each cabin could hold 19 guests plus the conductor. The cabin also provided comfortable seating, though most stood to enjoy the view.

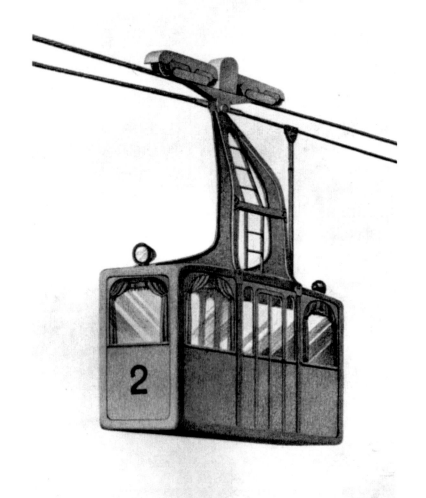

A Bleichert passenger carrier

Operations

For safe operations, both cabins and their conductors were in constant communication with one another and the Valley Station where the system operator was housed in a control room with a head-on view of the line. The cabin conductors and operator could activate warning bells at each other's control panels. Any of the conductor or operator positions could bring the ropeway to an emergency-stop. In the unlikely event that the hauling rope snapped, full-brakes would be automatically applied.

System speed was controlled by the operator, and cabin conductors did not have alternate throttles. The operator could also select from dead-slow and emergency-stop braking systems. When either station was approached by a cabin, a bell and indicator alerted the operator, and a relay automatically reduced the speed of the hauling rope. If an operator failed to manually guide the cabin into the station, the cabin would automatically pause at the station entrance. When a cabin was brought in and reached a station platform, a mechanism was tripped, halting the cabin. This made a cabin collision with a station highly unlikely, if not impossible.

Should the primary hauling rope fail, the cabin would halt and anchor itself to the track rope. Once the runner assembly engaged the secondary hauling rope, the cabin would release the track rope, and the runner assembly would move freely again.

Wind speeds are sampled at towers along the line and communicated to the operator by bell and indicator. High gusts meant the line was slowed or stopped. A system was also available that, depending on the maximum wind speed that was selected, would automatically stop the line during gusts or storms.

The Zugspitzbahn climbed 5,187 feet over its 11,089-foot length. The Valley Station and Mountain Station were 9,843 feet apart, and the trip between them took 16 minutes.

Construction

In order to access the new work-site at Obermoos, a road was cut through the forest from the village of Ehrwald. Along the ropeway's alignment, difficult dynamiting was executed by the Innsbruck-based construction firm *Kleiner*, and foundations were formed for the towers and both stations. This work was accomplished under extreme climatic and technical difficulties.

During manufacturing, major ropeway components were tested to their limits. For example, at *Adolf Bleichert & Co.*'s Eutritzsch factory, the cabin runner assembly clutch was tested for its ability to hold-fast a fully loaded cabin at track rope angles up to 90°, and under icy conditions. To facilitate construction of the main line, the company erected a simple and temporary wire ropeway along the steep Zugspitzbahn alignment. This temporary ropeway hauled explosives, worker provisions, and sand and water for mixing concrete at the various work sites.

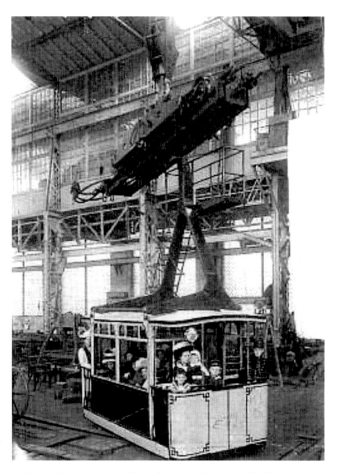
Load testing at Bleichert's Eutritzsch factory.

To the left of this main system tower is the temporary single-rope system used during construction of the main line.

Construction camps were established at tower and station sites. These camps had huts and kitchens for up to 400 workers. Of the system's construction tasks, the most challenging was the transport and installation of the track rope as its massive spool was awkward and difficult to handle.

The contract had stipulated that local sources be used where feasible. Therefore, the project included *St. Eqydier Iron & Steel Works* of Austria; *Felten & Guillaume Carlswerk* as supplier of the ropes; and tower steel components came from *Machine Works Simmering* of Vienna. Both the lower and upper stations were built by *Kerle Construction* of Reutte, and *Ed. Ast. & Co.* provided concrete, and, as already noted, the challenging explosive work, foundation, and masonry work was completed by *Kleiner*. Electrical components for the ropeway were provided by *Siemens-Schuckert*. Most of these firms were Austrian, thus providing the jobs and industrial offsets that Vienna desired when underwriting the project.

Construction. Note the giant spool of wire rope.

In 1926, the Reutte Power Works completed an extension of transmission lines from its generation plant to the Obermoos ropeway site. By March of that year, test trips on the system had started. Anticipation among locals ran high.

The President of Austria, Dr. Michael Hainisch, visited the Valley Station on June 14, 1926. He inspected the ropeway facilities, and, despite pouring rain, indulged in a ceremonial ride.

Presidential ride.

On July 5, 1926, the Zugspitzbahn was officially blessed, opening in the presence of the Minister of Transportation and 40 other various ministers from Vienna. The Innsbrucker News reported: "At the end of the festivities, the members of the federal government and the state representatives, together with District Magistrate Stumpf, entered the first carrier for the trip up the Zugspitze. All participants were very enthused about the impressive design and the wonderful views of the Ehrwald Valley."

Conclusion

In 1933, Germany imposed sanctions on Austria. This severely reduced traffic on the Zugspitzbahn and the old funicular, and, since both systems depended on cross-border tourism, profits dried up. Weighed down by 7-million-Schillings in debt from the project, the underwriters—the Tyrolean State, the village, and the Reutte Power Works—had to make large debt service payments. The village eventually forced repayment of its debt share on to the Reutte Power Works, severely hobbling the finances of this utility for years to come.

In 1937, shares in the wire ropeway were merged with those of the *Bavarian Zugspitze Co.* With the Second World War and the absorption of Austria by Germany, the ropeway was under-used and neglected. In 1945, American fighter-bombers attacked both stations of the Zugspitzbahn. The Valley Station and hotel were destroyed by bombs, and the wooden portion of the Mountain Station was shredded by gunfire. The hauling rope was also damaged by shrapnel and direct gunfire, though, as the line was not running at the time of the attack, there were no casualties. On June 1, 1945, American troops helped repair the wire ropeway.

In 1952, the Zugspitzbahn was refurbished with lighter cabins that held up to 23 passengers, and, in 1955, the wire ropeway came under Austrian administration. With this transfer, updates were made to the line, bringing it to a travel speed of 16.4 feet per second.

In 1958, the *Tiroler Zugspitzbahn Co.* was incorporated and headquartered in Ehrwald. The State of Tyrol hired *Pohlig Co.* of Köln to convert the line to four-cabin operation with a new stopover station at the Wienerneustadter hut. By 1964, no components of the original wire ropeway were in use, and a new summit ropeway began operating.

In its modern form, the Zugspitzbahn continues operation to this day, providing high-speed access to the summit in sleek modern cabins. There is a museum at the Mountain Station that commemorates the long history of conquering the Zugspitze with wire rope.

PREDIGTSTUHLBAHN, GERMANY

The Mountain Station at Predigtstuhl's peak.

Bad Reichenhall is located 1,552 feet above sea level and near the Austrian border, southwest of Salzburg. Inhabited since the Bronze Age, the area organized salt production around 450 B.C., and began exporting brine via an innovative pipeline in the early 1600s. In the 19th century, Reichenhall became known for its health resort and, from 1890, *Bad* (bath) was added to its name. Bad Reichenhall became the administrative center of the Berchtesgadener Land district of Upper Bavaria and, suffering under the hyperinflation of post-WWI Germany, sought to expand its tourism industry by offering hikers and outdoor adventurers' access to the natural wonders that surrounded it.

On June 16, 1927, with financing from Alois Seethaler of the Hotel Axelmannstein and spa director Josef Niedermeier, Bad Reichenhall placed an order (No. 3060) with *Adolf Bleichert & Co.*'s passenger wire ropeway division. Bleichert's engineers got to work.

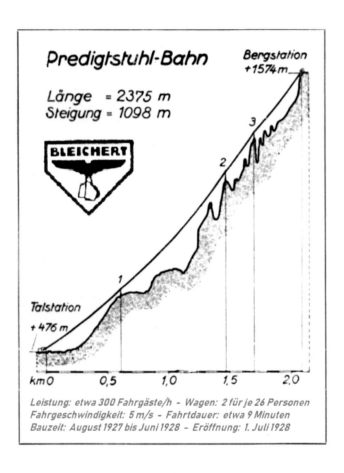

The *Predigtstuhlbahn* utilizes the Bleichert-Zuegg system to operate two carriers between a Valley Station and a Mountain Station. The carriers each have a capacity of 25 passengers and 1 conductor and can transport 300 passengers per hour over the 4.9-mile ropeway. Controlled by an operator located in the Mountain Station's control room, the carriers travel between 11.5 and 16.4 feet per second (an average of 11.2 miles per hour) and can move between stations in 9 minutes. The ropeway is supported by three steel-reinforced concrete pylon-shaped towers that jut from the steep mountain face.

Departing the Valley Station, the 75% slope of the line climbs over the Saalach River, a roadway on the opposite bank, and then over the mountain forest. Climbing the mountain, the carrier approaches Tower I.

Jutting from the mountainside, Tower I is 72 feet high. After clacking over the tower's rails, the carrier leaps a 3,248-foot span to Tower II that is perched on the mountain's shoulder.

Tower II is 105 feet high and, after a shorter span to the 30-foot-high Tower III, the carrier slows as it approaches the Mountain Station.

Tower I

Tower II

Construction

Though prime contractor *Adolf Bleichert & Co.* designed the *Predigtstuhlbahn* and manufactured the system's carriers, running gear, clutches, tower hardware, tensioning devices, control systems, and temporary construction ropeway, the core competencies of other German firms were enlisted for the project.

Tower I under construction

Manufacture of the wire ropes was sub-contracted to *Westfälische Drahtindustrie* (Hamm) and would total some 113 tons of spiral (hauling) and full-locked (track) types. *Hochtief AG* (Munich) was assigned the construction of the system's three concrete and steel support towers as well as the Valley and Mountain Stations. *Garbe-Lahmeyer AG* (Aachen) provided the ropeway's 150 horsepower electric traction motor and other electrical equipment, while *Güldner-Motoren GmbH* (Aschaffenburg) built the diesel engines and dynamos for generation of electricity. *Baer & Derigs GmbH* (Munich) designed and installed the hot water system for both stations, while *Michael Brandner* company (Bad Reichenhall) painted the stations and restaurant, and *Georg Kammel* company designed and built the custom furniture and other carpentry.

The manufacture of components and construction at the site of the *Predigtstuhlbahn* began in August of 1927. On the mountainside, 22 tons of dynamite blasted 141,000 cubic feet of rock, clearing and leveling the tower locations. A temporary construction ropeway was installed and began to move materiel and workers up the *Predigtstuhl*.

Workers install the track rope.

Construction workers ride the temporary construction ropeway.

The Mountain Station under construction.

Spools of wire rope.

Workers manhandle equipment.

Built at Bleichert's Leipzig-Gohlis factory, the *Predigtstuhlbahn*'s carriers have iron wheel bogeys, hangers, and frames, lightweight aluminum body panels, and Perspex (cast acrylic sheet) windows. Innovative dampers minimize swinging of the carrier after passing a tower. The dodecagonal shape and signal red paintjob of the carriers have become iconic, and they have proven to be durable and well behaved in the wind. Bleichert named their unique carriers 'Pavilion'.

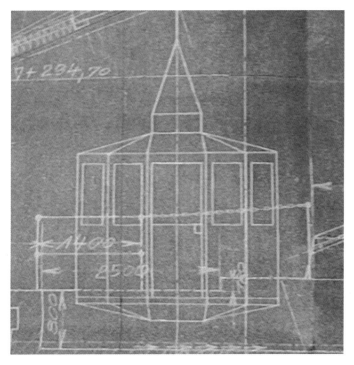

Pavilion carrier schematic.

Carrier in the Valley Station.

Carrier flying from the mountain mist.

Stations

Both the Valley and Mountain Stations are designed in the style of "New Objectivity" (*Neuen Sachlichkeit*), an interwar neo-realistic school of architecture that broke with Art Nouveau and sought modern functional design. Further, the station design was influenced by *Heimatstil*; 'Swiss chalet style'.

The Valley Station houses the system's tensioning system that consists of two 57-ton steel encased concrete weights suspended in 40-foot deep shafts. These weights maintain tension on the system's looped track ropes and minimize slack in the hauling rope.

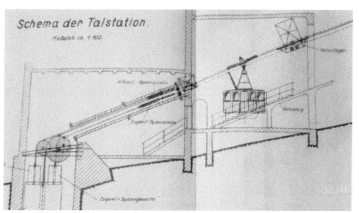

Valley Station schematic.

A carrier departs the Valley Station.

Valley Station exterior.

The Mountain Station anchors the spring-cushioned flywheels of the track and hauling ropes, a total anchorage load of 166 tons. The machine room houses the traction motor, water heater, generator, and electrical works, and the operators control room overlooks the carrier bay.

Operators control panel.

Ninety-five percent of the control room's equipment is original, and includes markers that indicate carrier positions on the ropeway, an important feature in fog, as well as a line speed shift with ten positions. Telephone calls and carrier commands are transmitted through the ropes.

Traction motor reduction gear and flywheel.

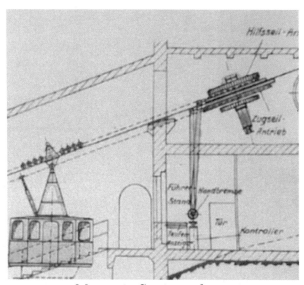
Mountain Station schematic.

Once at the Mountain Station, passengers step outside to a sweeping Alpine panorama and clean mountain air. An open-air café terrace invites one to lounge on sofas and chairs while aromas waft from the restaurant, hinting at local delicacies served on white linens and china. For the adventurous, trailheads beckon.

Walking along the main peak trail for some 15 minutes, passengers reach the rustic *Schlegelmulde* Alpine Hut, which offers a sun terrace, deckchairs, and warming room. Here, passengers can enjoy regional delicacies and Bavarian ambiance, absorbing the snow-covered winter landscape, the colorful flora of spring, the bright sunshine of summer, or the crisp chilly autumn breeze. From the hut, passengers can set off to discover more challenging trails that crisscross the *Predigtstuhl* and surrounding peaks.

Diesel engine and dynamo.

90th Anniversary

The 90th Anniversary of the *Predigtstuhlbahn* was celebrated on July 1, 2018, with events at both the Valley and Mountain Stations, and a guest list that included local and national administrators, businesspeople, politicians, as well as descendants of Adolf Bleichert.

On the occasion of the anniversary, the *Predigtstuhlbahn*'s carriers are estimated to have traveled over one million miles, all on the original wire ropes and using the original equipment. The *Predigtstuhlbahn* is a listed Historic Monument with respective protections, and continues its journey from the past into the future.

TABLE MOUNTAIN AERIAL CABLEWAY, SOUTH AFRICA

Cape Town was founded in 1652 and is South Africa's oldest city. Up against the Cape Fold Range, and nestled between the South Atlantic and Table Mountain's shoulder, Cape Town was settled by the Dutch East India Company as a way-station for ships traveling around the Cape of Good Hope and on to the Dutch East Indies.

Dutch ships in the waters off Cape Town, the settlement and Table Mountain flanked by Devil's Peak and Lion's Head Peak in the background, late 17th century.

Cape Town is located in one of Earth's six floral kingdoms, the Cape Floristic Region, home to endemic fauna and flora. Recognized as a UNESCO World Heritage Site and one of the New7Wonders of Nature, the Cape Floristic Region includes Table Mountain National Park, as well as seven other protected areas around Cape Town. Since Cape Town's founding, Table Mountain and its unique environment have enticed the adventurous to climb and explore.

Table Mountain could only be climbed by foot in the 1870s. A funicular was proposed and approved, but the First Anglo-Boer War interrupted plans in 1880. In 1912 it was decided that a rack railway would be constructed. However, the outbreak of World War I in 1914 meant this project got sidelined as well.

The question of how best to scale Table Mountain came up again in 1926, and Norwegian engineer Trygve Stromsoe proposed to the Cape Town City Council an aerial wire ropeway, a cableway. Despite high cost, some £60,000 of the day (~15 million 2019 USD), the citizens of Cape Town voted in favor of a cableway. The Table Mountain Aerial Cableway Company (TMACC) was formed to finance the project.

The TMACC board comprised businesspersons Sir Alfred Hennessy, Sir David Graaff, Sir Ernest Oppenheimer, as well as the engineer, Stromsoe.

On November 16, 1926, TMACC contracted with *Adolf Bleichert & Co.* to build the Table Mountain Aerial Cableway. Bleichert's engineers and production halls got to work.

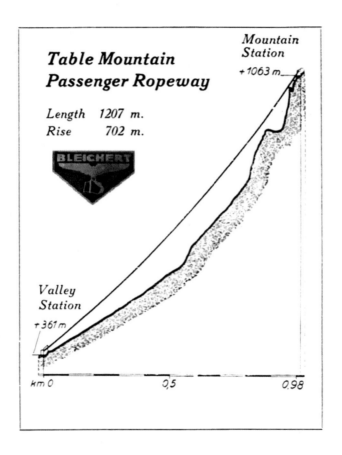

De/ign

For Table Mountain Aerial Cableway, Bleichert delivered two of its 'Rectangular' cabins to travel between a Lower Station and an Upper Station. Made of corrugated iron sheet panels and iron frame and hanger, the cabins had a maximum capacity of 24 passengers and 1 conductor each. With a line speed of 11.2 miles per hour, the cabins moved at 16.4 feet per second, and could climb Table Mountain in under 5 minutes. Communication with the control room is by telephone signal transmitted via the hauling rope.

Bleichert's 'Rectangular' cabin at Table Mountain, 1929.

There is a 3,960 foot/0.75-mile rope span between stations with no support towers. Bleichert delivered its spiral track and locked hauling wire ropes which weighed some 18 tons combined.

Stations

Local architects Walgate & Ellsworth under specifications outlined by Bleichert designed the Lower and Upper Stations. Construction of the station buildings was subcontracted to local industry, with Bleichert manufacturing all system components at its German factories, and shipping them to South Africa. In order to move materials and workforce, Bleichert installed a temporary ropeway at the Lower Station worksite and up to Africa Ledge, location of the Upper Station. When the buildings were done, Bleichert's technicians installed and tested all the cableway's equipment. Generally, both stations housed cableway components, electricity generators, cabin platforms, and other customer and engineering support facilities.

The Lower Station is at 1,190 feet above sea level. It is located on Tafelberg (Table Mountain) Road inside Table Mountain National Park in the Garden neighborhood of Cape Town, some 15-minutes from the city's downtown and waterfront. The Lower Station houses dual cabin platforms, sheave wheels, and two 149-ton rope-tensioning counterweights slung in 40-foot-deep shafts. The building also has administrative and facility spaces, as well as a visitor center and a shop.

Lower Station in 1940...

...and, in 2014.

A cabin emerges from the Lower Station. It begins a steep climb that matches that of the mountain's sloped granite skirt. The ropes run parallel to India Ravine and fly over India Venster trail up to where it meets Table Mountain Lower Station trail. Table Mountain then rears suddenly and its geology changes from granite to a near vertical sandstone plateau. The cabin ascends the wall of the Western Table, gliding high above a bowl formed between Africa Ledge and Kloof Corner Ridge. The cabin gets slammed by a strong gust off the Atlantic Ocean.

The cabin sways and creaks, and the vibrating wire rope bounces. Passengers laugh nervously. Ahead, Upper Station stands on the precipice of Africa Ledge. The cabin approaches the station's arched mouth and slows to meet the platform.

The Upper Station sits at 3,500 feet above sea level, some 2,310 feet higher than the Lower Station. A veneer of native stone blends the building into the surrounding geology. Upper Station houses the electric primary and auxiliary rope drives, transmissions, sheave wheels, a control room, docking rails, and a single movable cabin platform that can access either track rope. The control room provided a clear view down the line, had a back-up handbraking system, line throttle, indicator panels, signal bells, a telephone system for communicating with the Lower Station and cabins, as well as access to the machine and rope spaces

View down line from the operator's cab.

The cabin parks and passengers disembark, exiting Upper Station on to a sweeping paved patio. There is a mailbox that offers a unique Table Mountain postmark, and trailheads that promise ways to the Table Mountain Café, memorials, and observation points around the Western, Central, and Eastern Table. Standing at the station, the view includes Lion's Head Peak, Cape Town, Table Bay, and Robben Island, as well as the flat expanse of the Table with Devil's Peak and the Atlantic seaboard to the south and west. Visitor amenities include audio or guided walking tours, the Table Mountain Café (and, nowadays, a wi-fi lounge), historical registered structure, and various observation points.

Upper Station.

After two years of challenging construction, on October 4, 1929, Cape Town's Mayor Rev. A. J. S. Lewis opened the Table Mountain Aerial Cableway to its citizens and the world.

TMAC Updates

The Table Mountain Aerial Cableway has been updated four times since opening day: in 1958, 1967, 1974, and, most extensively, in 1997. The 1958 update replaced the original Bleichert 25-passenger cabins with new all-metal ones that held 27 passengers. The original wire ropes made by Bleichert were changed out, too, with new ones made in South Africa. The 1958 cabins ran until 1966.

The 1967 update included new cabins that held 28 passengers, and the electric drive motors and transmission were replaced in the Upper Station. Further, the ropes were lowered to the ground and electro-magnetically inspected.

The 1974 update replaced the 1967 cabins with newer, lighter ones, though cabin capacity remained the same at 28 passengers.

1958 and 1974 cabins.

The most ambitious and expensive update occurred in 1997. Under new ownership and contending with six-hour wait times at the busy cableway, the TMACC embarked on extensive modernization of the system. An R87 million (about 10 million 2019 US dollars) contract was signed with ropeway engineering firm Garaventa AG of Switzerland for installation of a Rotair-car System at Table Mountain.

Table Mountain Aerial Cableway's rope layout was changed from the original Bleichert bi-rope to Garaventa's modern tri-rope type. The tri-rope consisted of dual full-locked coil 238-ton rated track ropes and 75-ton rated haul and heel ropes. The new system located each cabin at the opposite end of the haul and heel ropes for counterbalance, operating in a jig-back system that harkens to Bleichert-Zuegg.

The system's dual track ropes could carry heavier loads while providing lateral separation that countered cabin sway in windy conditions. Primary control of the line was changed from the Upper to the Lower Station, and new rope anchors were installed consisting of 9-foot concrete bollards with 45-foot long steel anchors driven into the rockface. With a new tensioning system, counterweights were reduced to 134 tons. Lastly, the 1997 update included new cabins.

Rotairs

'Rotair' cabins were purchased from Swiss manufacturer CWA, now a subsidiary of the Doppelmayr-Garaventa Group. Rotair cabins were known for their use at Mt. Titlis, Engelberg, Switzerland, and, Mt. San Jacinto, Palm Springs, USA. The cabins utilize aircraft grade materials and are round, making them less susceptible to winds. The cabins can take on 793 gallons of fresh water in a tank beneath their floor. The water acts as ballast against high winds, expanding the cableway's operating envelope to 37 mph gusts, and can be transferred to a reservoir at the Upper Station for use at the restaurant and other mountaintop facilities.

A Rotair cabin holds 64 passengers plus one cabin master, or, alternatively, the cabin can haul 5 tons of payload up the mountain, supplying the restaurant, shops, and café. Rotairs are named for their rotating floor that provides passengers with panoramic views. A waste tank can be slung beneath the cabin to bring sewage down to the Lower Station for dumping into Cape Town's municipal system.

Each cabin and its hanging gear weigh approximately 11 tons, and is suspended from the track ropes with a 24-wheel carriage and hanger. This running gear sports special shafts equipped with dampers that ensure optimal contact between the wheels and rope, further reducing cabin swing.

Power & Brakes

A 540-kilowatt alternating current induction motor was installed in 2013, and is controlled by four variable frequency drives. If municipal electricity is cut, the cable has two back-up hydraulic systems that can power the cableway. The first is powered by a Volvo engine and the second with one from Deutz.

Braking the line is accomplished with the service brake that affects the gearbox shaft of the haul rope, and there is an emergency brake that can stop the haul rope drive wheel. Lastly, a cabin master can brake and lock a cabin independent of station control.

The new Rotair-car System allowed a line speed of 33 feet per second, some 19 feet faster than before. The new capacity and speed provided by the 1997 update enabled Table Mountain Aerial

Cableway to carry 800 passengers per hour in trips of under five-minutes between stations.

Both stations were modified in 1997 to accommodate new machinery, the larger cabins, and were generally refurbished. Cabins from 1958 and 1974 were converted to information and ticket booths and were installed near the Lower Station entrance.

A Rotair cabin approaches the Upper Station. Lion's Peak, Cape Town, and Robben Island are in the background.

The cableway was closed in 1996 for construction, then reopened on October 4, 1997, the 68th Anniversary of the cableway's opening day in 1929. Ridership soared with the improvements. In the high season, upwards of 10,000 people ride the cableway per day, and 2018 saw record overall ridership of 855,000.

Conclusion

The Table Mountain Aerial Cableway is one of Africa's greatest tourist attractions, and celebrates its 90th Anniversary on October 4, 2019. The Table Mountain Aerial Cableway has a rich engineering history, and continues its endless journey at Cape Town's Table Mountain.

The Table Mountain Aerial Cableway operates in one of the most unique natural places on Earth. The cableway has a rich engineering history and a bright future.

END OF AN ERA

The Great Depression soon threw the world in to economic chaos, and market demand fell precipitously. Sales at *Adolf Bleichert & Co.* tumbled and many of its competitors went bankrupt. In 1929, the company recorded just 49 orders, followed by 30 in 1930, and 11 in 1931.

In 1926, it was decided that, for reasons of succession, *Adolf Bleichert & Co.* would become public. This added the acronym 'AG' (Aktiengesellschaft, a public company) to the company name, and 4,000 shares were issued, though the Bleichert brothers owned 100 percent of said shares, with Max holding three more shares than Paul.

In 1927, Paul von Bleichert separated from the company for health reasons, selling his 1,997 shares in a January 10 secret sale to *Felten & Guilleaume Carlswerk*, manufacturer of wire ropes. This reduced family ownership in *Adolf Bleichert & Co. AG* to just over 50%.

With the July 7, 1931 collapse of the German banking system, Max von Bleichert lacked the equity to maintain liquidity at the firm. After a last-ditch effort by Max to raise funds by selling his personal art collection, on April 4, 1932, *Adolf Bleichert & Co.* filed for bankruptcy. A last-minute plan was engineered, and *Felten & Guilleaume Carlswerk* stepped in, reorganizing the company. Its successor, *Bleichert-Transportanlagen GmbH*, was incorporated on June 28, 1932 to carry on the firms work. *Bleichert-Transportanlagen GmbH* also became sole shareholder of Adolf *Bleichert & Co. Drahtseilbahn GmbH*, the people-mover manufacturing entity. *Bleichert-Kabelbagger GmbH*—the wire rope crane division—became an independent entity, though also declared bankruptcy on July 4, 1932.

On September 18, 1938, following eye surgery in Davos, Switzerland, Paul von Bleichert died unexpectedly of pneumonia. He was buried at Gohlis Sudfriedhof (South Cemetery).

In Germany, Hitler's National Socialists had come to power.

World War II and the Iron Curtain

No longer under Bleichert family control, the *Bleichert-Transportanlagen GmbH* factory continued to produce during the Second World War. Targeted by Royal Air Force bombing raids on Leipzig, both the factory and adjacent family residence (Villa Hilda) suffered damage from bomber raids, especially during those of December 4, 1943 and February 27, 1945.

With the defeat of Nazi Germany, Leipzig fell on the eastern side of the Iron Curtain. *Bleichert-Transportanlagen GmbH* was taken over by the Occupying Power, the Soviet Union, and renamed *SAG Bleichert*.

On January 19, 1947, Max von Bleichert died of natural causes. He was buried at the Stadtfriedhof (Municipal Cemetery) in Goettingen, West Germany on January 21, 1947.

SAG Bleichert continued to manufacture conveyance systems primarily for delivery to the Soviet Union as war reparations. During Communist administration, employment increased by almost five times and production increased 1000%. Now also manufacturing the 'Karlik' line of battery-powered tractors (These machines were often labeled as produced in Poland or the Soviet Union), the company built an additional 52 aerial wire ropeway installations in the Eastern Bloc. However, the Soviet Union moved most wire ropeway production to within its own borders, quickly becoming the primary supplier of wire ropeways for its satellite states. Thereafter, *SAG Bleichert* focused on soft-coal conveyance systems.

In 1954, *SAG Bleichert* was transferred to the German Democratic Republic (East Germany), and was renamed *VEB Bleichert*. Soon thereafter, the firm was continued under the name *VEB Transportanlagenfabrik Bleichert Leipzig*. In 1955, the company name changed again to *VEB Schwermaschinenbau Verlade- und Transportanlagen Leipzig vorm.Bleichert*. By 1959, the last reference to the original family business disappears, as 'vorm.Bleichert' is dropped from the firm's name. Between 1962 and 1985, this entity went through several iterations. However, by 1991, the company had been privatized and entered liquidation, halting production of cranes, conveyance, and pit mining equipment—thus concluding the history of the oldest and largest wire ropeway manufacturer of the world.

Bleichert-built wire ropeways still exist and operate in Germany and around the world. It is also likely that many more Bleichert Systems are located and possibly continue to operate in the successor states of the Soviet Union.

Though the exterior/shell of the Leipzig-Gohlis factory is protected as a national historic site, in 2009, *Gohliser Hoefe GmbH & Co.KG of the CG Gruppe* bought the factory with the intent of redevelopment as office and residential space. As of 2016, this redevelopment progresses, and an architectural plan that respects the site's industrial history is being adhered to.

Villa Hilda, the historical Bleichert home adjacent to the factory, stands today as the Heinrich Budde House, and provides space for artists and as a community center. Its restoration continues.

Certain streets in Leipzig have returned to pre-Communist names, including 'Gohlis/Eutritzsch Bleichert Street,' 'Bleichert Street', and a commemorative street now exists in the national capital of Berlin.

Though several firms bear the Bleichert name, such as *Bleichert-Automation GmbH* and its American subsidiary, *Bleichert, Inc.* offering automation and conveyor technology; and, *Bleichert Förderanlagen GmbH* offering automation, conveyor, and gantry systems, none of these entities are related to the original *Adolf Bleichert & Co.* Research has determined that Adolf George Bleichert—a younger brother of Max and Paul—lent his name to the firms. Companies operating today evolved from this enterprise.

SOURCES

90 years of Predigtstuhlbahn. HOW TO BUILD LEGENDS. Retrieved from www.predigtstuhlbahn.de

Adolf Bleichert & Co. (n.d.) "Die Zugspitzbahn". No. 579.

Bleichert Aerial Ropeways (Bi-Cables). Adolf Bleichert & Co. A.G. Leipzig-Gohlis, Germany.

Bleichert System Wire Ropeway. Deutsches Kolonial Lexikon, Vol. II & III. 1920.

Chilecito Tours. "Aéreocarril." 2005.

Customer Register of Bleichert Transportanlagen GmbH. State Archives of Saxony at Leipzig, Germany.

Bleichert Transportanlagen: Customer Register No.52 (1875-1923). "Chilecito." State Archive of Saxony. Leipzig, Germany.

De Decker, Kris. "Aerial ropeways: auto cargo transport for a bargain". Low-tech magazine. January 2011.

"Die gigantische Österreichische Zugspitzbahn—"die kühnste und eindrucksvollste Seilschwebebahn der Welt" 1926—1991". Retrieved on June 1, 2009, from, http://www.seilbahngeschichte.de/erstetzb.htm

Dietrich, G., *Die Aufschliekung der Nickelerzlagerstätten in Neukaledonien*, Sonderabdruck aus der Zeitschrift des Vereines deutscher Ingenieure, Berlin 1907.

Dietrich, G. "Zeitschrift des Vereines Deutscher Ingenieure." No. 44, Nov. 3 1906; No. 45, Nov. 10 1906; No. 46, Nov. 17, 1906.

Hiscock, David. "Cape Town's Man-made Wonder: Table Mountain Cableway - The First 80 Years. " Table Mountain Aerial Cableway Company, 2010.

Hoetzel, Dr. Manfred. "Adolf Bleichert und sein Denkmal in Gohlis." Buergerverein Gohlis, 2000. Leipzig, Germany.

Keilhack, Jutta. "Der Sachsenkoeing kam zum Fruehstueck." Harzburger Zeitung, 1989. Harzburger, Germany.

Koehler, Dr. G.W. Adolf Bleichert & Co. "Retrospect at the Occasion of its Fiftieth Anniversary on July 1, 1924." Polytechnic Institute. Darmstadt, Germany.

La Nacion. "El cablecarril—Chilecito se cotiza tan alto como el oro de sus sierras." 1996. Buenos Aires, Argentina.

Minister of Public Works. "Testimonial by the Government of the Republic of Argentina." Oct. 7, 1907. Buenos Aires, Argentina.

Mkumbara-Shume: About the Wire Rope Way. Usambara Post. August 1909.

Mkumbara Wire Ropeway. Deutsches Kolonial-Lexikon, Vol. II & III. 1920.

Nationsencyclopedia.com "French Pacific Dependencies—New Caledonia."

Neuman, Holger. "125 Jahre Fa. Bleichert." Werkbahnreport, 1999. Historiche Feldbahn Dresden. Dresden, Germany.

Productivity and Costs for Skyline Logging. Ole-Meiludie, Raphael and Abeli, Wilbald & Silayo, Dos-Santos A. & Shemwetta, George. Wood for Africa Forest Engineering Conference, July 2002.

Schmaedicke, Dr. Juergen. "Bildungswerk der Saechsischen Wirtschaft." Kalendar, 1995. Leipzig, Germany.

Tanzania. InfoPlease.com, almanac and atlas.

Tiroler Zugspitzbahn (2008). "With the best views..." Retrieved on June 1, 2009, from, www.zugspitze.at

Trennert, Robert A. "Riding the High Wire: Aerial Tramways in the West." University Press of Colorado, 2001. Boulder, Co., USA.

UNU.edu, "Nickel Mining on New Caledonia."

Verhandlungen des Vereins zur Befoerderung des Gewerbefleisses. Issue 6. Publisher Leonhard Simion Nf., Berlin, 1911.

Wettich, Hans, Engineer. *Die Entwicklung Usambaras unter dem Einfluss der ostafrikanischen Nordbahn und ihrer privaten Zweigbahnen mit besonderer Beruecksichtigung der Drahtseilbahn Mkumbara/Neu-Hornow.* "Verhandlungen des Vereins zur Befoerderung des Gewerbefleisses." Adolf Bleichert & Co. Berlin. Issue 6. 1911.

Wire Ropeway at the Schumewald. Usambara Post. February 1911.

Wire Ropeway to the Shume Forest. Deutsche Colonialzeitung. 1909. Pg. 583

Wolf, Heinz. "Der Erfinder Adolf Bleichert (1845-1901)." Motor im Schnee—Historische Ausgabe. Iffezheim, Germany.

Wolf, Heinz. "Die Bleichert-Montanbahn von Chilecito." Motor im Schnee—Historische Ausgabe. Iffezheim, Germany.

* German-to-English source translation by Rolf von Bleichert
* Spanish-to-English source translation by Peter von Bleichert
* Thanks to Hartmut von Bleichert

Made in the USA
Columbia, SC
14 July 2022